Whoops!/WPPSS

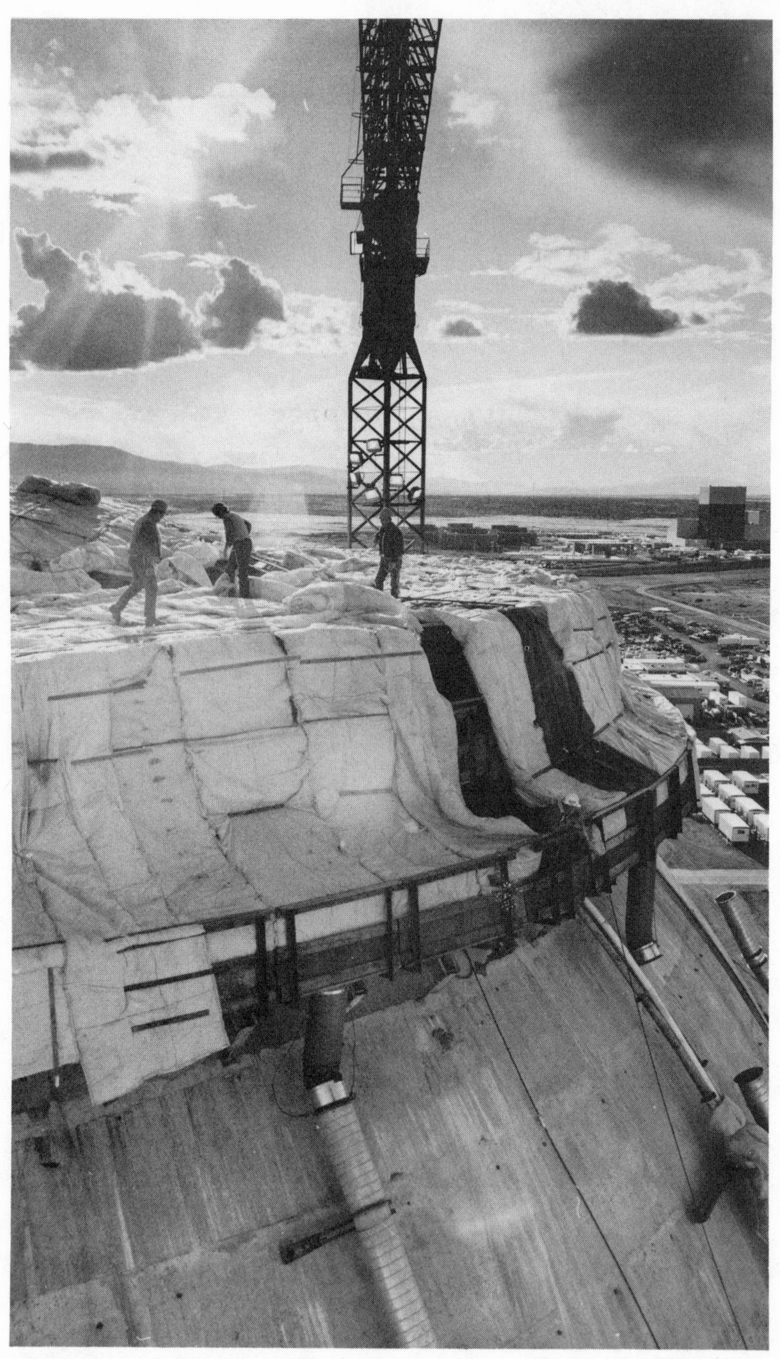

Whoops! / WPPSS

Washington Public Power Supply System Nuclear Plants

by
David Myhra

McFarland & Company, Inc., Publishers
Jefferson, North Carolina, and London

Frontispiece

Insulation-covered scaffolding permitted concrete work on WNP 1 containment building to continue through the winter months, ca. 1982.

Library of Congress Cataloging in Publication Data

Myhra, David.
Whoops!/WPPSS : Washington Public Power Supply System nuclear plants.

Bibliography: p.
Includes index.
1. Washington Public Power Supply System — Finance.
2. Electric utilities — Washington (State) — Finance.
3. Atomic power-plants — Washington (State) — Design and construction — Finance. 4. Default (Finance) — Washington (State). I. Title.
HD9685.U7W345 1984 363.6′2′09797 84-42601

ISBN 0-89950-128-1

Manufactured in the United States of America.

McFarland Box 611 Jefferson NC 28640

To Julie

...The Participant shall make the payments to be made to Supply System under this agreement whether or not any of the projects are completed, operable, or operating and not withstanding the suspension, interruption, interference, reduction or curtailment of the output of either project for any reason whatsoever in whole or in part. Such payments shall not be subject to any reduction, whether by offset or otherwise, and shall not be conditioned upon the performance or non-performance by Supply System or any other participant or entity under this or any other agreement or instrument, the remedy for any non-performance being limited to mandamus, specific performance or other legal or equitable remedy...

—From the official bond resolution, adopted by Washington Public Power Supply System on February 23, 1977, providing for the issuance of revenue bonds.

Table of Contents

Preface

Whoops!/WPPSS is the story of Washington Public Power Supply System's now terminated Nuclear Project 4 and 5 and its default on $2.25 billion worth of tax-exempt bonds to an estimated 100,000 bondholders.

There is a temptation to see this unprecedented event as a story of villains—high interest rates, inflation, an ill-equipped Board of Directors, the 88 Participants, the Washington Supreme Court, greedy contractors and labor unions, the nuclear electric plant accident at Three Mile Island, the investment community, the U.S. Nuclear Regulatory Commission, and the Bonneville Power Administration. Before reading conclusions into this story, one ought to examine all the evidence. What happened to WPPSS and its Pacific Northwest Participants in the ill-fated nuclear projects can be demonstrated by a common occurrence that might be called the "vector principle." A boat sets off for opposite shore of a river but because of various unconsidered currents ends up several miles downstream. The WPPSS program for five nuclear plants in the early 1970s was thought to be a reasonably obtainable, rational goal but it failed nonetheless because it did not take fully into account the currents of nuclear power plant economics and construction complexities of building even one or two plants at a time let alone five simultaneously.

Blindness toward these realities was heightened by the historical interaction of WPPSS in the Federal Columbia River Power System—the discouragement of public participation and honest reporting of the need for electric power by the Bonneville Power Administration. Especially remarkable is the extent to which WPPSS was allowed to pursue its single-minded strategy over economic realities before it was too late to avoid the largest municipal default in the nation's history.

My research efforts have benefitted from a series of articles on the WPPSS "power program gone awry" written by *The Oregonian* (Portland) reporters Donald J. Sorensen and John Hayes.

The literary effort was aided by Sherrell Varner, who read successive stages of the manuscript and gave needed editorial criticism.

To avoid overwhelming the reader with footnotes, I have generally attributed only direct quotes. In most cases, the background notes and bibliography should serve as a guide for those wishing to do further study.

Naples, Florida *D.M.*
May, 1984

Introduction

Since July 4, 1960, when Commonwealth Edison's Dresden nuclear electric power plant #1, near Morris, Illinois, began the first commercial production of nuclear electric power, there have been at least 236 nuclear plants in various stages of planning or construction in the United States. Eighty-two nuclear plants are currently in operation, and some 48 are in various stages of construction around the country. These plants have had their difficulties—safety problems, even accidents, construction delays, cost overruns—but none has had bigger problems than the Washington Public Power Supply System's (WPPSS) five nuclear electric plants.

Now known derisively as "Whoops!", WPPSS has been the most expensive and most mismanaged nuclear electric power plant construction effort in the United States. Spiraling cost overruns plagued the program from the beginning. However, the real depth of WPPSS' problems became apparent only after management's decision in January 1982 to terminate two of the five plants after having spent $2.25 billion raised by the sale of tax-exempt municipal revenue bonds. The bonds on these two plants—known simply as Washington Nuclear Plants (WNP) 4 and 5—were guaranteed by 88 Participants, small municipalities and public utility districts in the Pacific Northwest. Under terms of individual "take or pay" contracts with WPPSS, each Participant had agreed to be responsible for its share of the debt, whether or not the projects were completed and whether or not they were operable after completion.

After terminating WNP 4 and 5, WPPSS turned to its 88 partners for funds to begin paying the debt service (that is, the interest) on the revenue bonds. To WPPSS' horror, most of the "partners" refused to pay. They claimed that they had not had the legal authority to enter into such agreements in the first place and that, therefore, the "take or pay" contracts were not valid. In addition, they said, the Bonneville Power Administration (BPA), backer of WPPSS WNP 1, 2, and 3, had threatened them with power shortages, brownouts, blackouts, and other difficulties if they didn't sign up with WPPSS to build WNP 4 and 5. This coercion, they said, also made their contracts with WPPSS invalid and therefore non-binding. WPPSS officials went to court, never believing for a moment that they would not be able to force the 88 partners to pay up. The issue moved quickly through the courts. On June 15, 1983, the Washington State Supreme Court ruled that the 88 Participants did not have the authority to enter into agreements with WPPSS and freed them from their obligations on the bonds. Shortly thereafter, WPPSS defaulted. It was the largest municipal bond default in U.S. history.

Since June 1982, there have been a number of efforts, both private and public, at the State and Federal levels, to aid WNP 4 and 5 bondholders. Early efforts led by Washington Governor John Spellman came to naught, and a bailout by the Federal government was always considered unlikely. Efforts on behalf of the bondholders by the Chemical Bank of New York, trustee of the tax-exempt bonds, have so far been unsuccessful, too. One plan suggested by the governors of Oregon and Washington in July 1983 which enjoyed a fair amount of acceptance for a time was to collect funds through a special charge on electricity transmitted or used in the Pacific Northwest. The special charge would cost a typical resident of a service area an extra 72¢ a month and would remain in effect for the life of the bonds, which in some cases is 35 years.

While others debate how the debt should be paid, if at all, a growing number of angry investors in WNP 4 and 5 are taking matters into their own hands and are pressing legal efforts to secure a portion of the assets and revenues of WNP 1, 2, and 3. Those plants are being financed by tax-exempt municipal bonds, too, but they are guaranteed by BPA, a Federal agency that markets all the electrical energy produced by federally owned hydroelectric projects on the Columbia and Snake rivers, including Hoover and Grand Coulee Dams. However, those three plants are not in much better shape: WNP 1 and 3 have been mothballed, perhaps indefinitely, while WPPSS struggles to get WNP 2 operational in the face of massive cost overruns and construction delays.

* * *

Headquartered in the small city of Richland, Washington, WPPSS is a curious creature. A joint venture of four municipalities and 19 electric utilities owned by various local governments, the forerunner to WPPSS got off to modest start in 1957 by building a single hydroelectric project at Packwood Lake, Washington. With a respectable track record and broad backing within the State, WPPSS was a natural choice in the early 1970s to undertake a new electric energy project. The project, as it turned out, was to finance, construct, and operate five nuclear electric power plants.

For years, the Pacific Northwest had thrived on inexpensive hydroelectric power. (Even today, customers there pay less than 2¢ per kilowatt-hour, about one-sixth the rate in New York City.) By the late 1960s, it became apparent that the bargain electric rates were boosting consumption so fast—about 7 percent a year—that hydroelectric power could not meet the Pacific Northwest's future needs. WPPSS initially decided to build three nuclear electric power plants, each of which could produce about 1,200,000 kilowatts (the number was increased to five later). All plants would be located in Washington State.

WPPSS contracted to supply power generated by these plants to its members as well as to other public electric utilities in Idaho, Montana, Oregon, and Utah. Then, on the strength of those contracts, WPPSS issued the first of what would be many bond issues that would eventually total $8.3 billion worth of tax-exempt municipal bonds. Interest rates were attractive and investors throughout the United States lapped them up. But what looked like five riskless ventures has turned into a financial and technological nightmare. The first power production, originally due in 1977, will not come until 1985—if then. Estimated costs of the five plants, initially set at $6.67 billion, had soared to $23.78 billion the last time anyone calculated. In January 1982, when WPPSS terminated WNP 4 and 5, the estimated cost of just these two plants had jumped from $3.5 billion to $11.8 billion.

With $6.8 billion in outstanding bonded debt ($4.5 on WNP 1, 2, and 3 and $2.25 on WNP 4 and 5), WPPSS is the nation's largest issuer of municipal tax-exempt bonds. The five unfinished nuclear electric plants carry the greatest long-term tax-exempt bond debt in history, even more than New York City owed when it was threatened with bankruptcy in 1965. Furthermore, WPPSS would have to raise an additional $12.5 billion just to complete the five plants. Now it appears that only one of the five nuclear plants will ever be completed.

Who will repay the $2.25 billion WPPSS borrowed for its now-terminated WNP 4 and 5? Why did the Participants go back on their promises? "We had to do it," says Major Ted Rakoski of McCleary, Washington. "There is no way for most cities to raise those funds." McCleary is a forest town of 1,500 people west of Olympia. Its share of the debt on the two canceled plants came to nearly $3 million, or $2,000 per capita. The share for Bonners Ferry, Idaho, with a population of 1,900, amounted to $4.3 million, about $2,263 per capita. The story is the same at the other municipalities and public utility districts WPPSS contracted with. All would have to consider going into bankruptcy, thus dimming their hopes of borrowing money in the future. One utility, in fact—the Orcas Power and Light Company, which serves several small communities north of Seattle—did file for protection from its creditors under Chapter 11 of the U.S. Bankruptcy Act, in December 1982. Interim Manager Bob Snow said the tiny company had been told by WPPSS that it owned $45 million on the two terminated power plants, with $2 million of that, about half of the utility's annual income, due in 1983. Paying would risk the collapse of Orcas Power and Light.

What went wrong to require such a drastic step as termination of not one, but two nuclear plants? Certainly other nuclear electric power plants have been canceled, though most of them were terminated while still on order, or at least before construction started. According to the Atomic Industrial Forum (AIF), an electric utility research and public relations organization based in Washington, D.C., no plant more than 5 percent completed had ever before been terminated. WPPSS stopped construction on WNP 4 when it was 23 percent complete and WNP 5 when it was 17 percent complete, after spending $2.25 billion.

Apologists say that "everything" went wrong—labor strife, regulatory delays, declining demand for electricity due to economic recession and conservation, and high interest rates. Still, such problems might be considered "par for the course" in the nuclear electric industry. Delays are not unusual, and plants have been "mothballed" from time to time. A classic case was the Arizona Public Service Company's one-year delay in 1979 in construction of its Palo Verde #1 project when it was 41 percent completed. The company blamed the action on high worker turnover and inclement weather. Work resumed in 1980, and the project is scheduled for commercial operation in December 1985, nearly nine years behind schedule.

Adverse financial conditions have hit other nuclear projects but they have still managed to move along. Some plants have survived in the face of problems that WPPSS has not had to contend with, for example, delays due to spirited action by environmentalists and antinuclear groups.

There has been some labor unrest at the WPPSS projects, but nothing of any great significance, and from the beginning WPPSS enjoyed easy access to the New York City bond markets. Until a so-called "taxpayer revolt" in November 1981 which threatened to strip them of the power to borrow money for the projects without voter approval, WPPSS officials had been raising all the money they needed, whenever they needed.

Critics blame management stupidity and bungling for most of WPPSS' problems. In addition, WPPSS has been plagued by lawyers who wrote contracts whose legality had never been tested in court, investment houses eager to sell bonds and earn commissions, and a Board of Directors composed for the most part of farmers, ranchers, and small businessmen who held part-time political office and were utterly lost in the megabucks world of nuclear electric economics. It appears that the WPPSS nuclear electric building program and its subsequent financial meltdown will provide a number of lessons.

First, WPPSS' lack of leadership and inability to carry out its program strongly supports the arguments of some critics of nuclear energy (and even some proponents) that only large organizations experienced in the construction and operation of nuclear plants should be allowed to tackle such a large, complex project as planning, building, and operating even one nuclear plant, let alone five.

Second, the financial collapse of WPPSS and the bondholders' lack of legal recourse cast serious doubts on the validity of so-called expert legal opinion as to the status of securities issued by organizations such as WPPSS. Bond sellers nationwide proclaimed the WPPSS bonds risk-free owing to the "take or pay" contracts between WPPSS and the 88 participating utilities. The refusal of WPPSS participants to pay their shares as promised suggests that one should not count on someone else's word—or even signature. Instead, the investment community, underwriters as well as buyers, should as part of their risk assessment look closely at the economics underlying a project and assign less importance to contracts for selling the power. In addition, investors interested in power projects of all kinds should look more closely at why the project is being undertaken, whether or not it makes sense, whether or not the forecasts for the need for power appear reasonable given the information that is available, and, finally, whether or not the project seems commercially viable.

1

The Setting

On September 26, 1963, President John F. Kennedy addressed a crowd of 40,000 well-wishers gathered at Hanford, Washington, for the groundbreaking of the Washington Public Power Supply System's (WPPSS) first nuclear project, the Hanford nuclear electric generating plant. Holding a uranium-tipped wand, Kennedy declared: "On this site ... a nation dedicated to living in peace is forging not a sword, but a plowshare, the Hanford steam electric generating plant." He summoned the Pacific Northwest and the nation to enter the nuclear age: "As it is well known here at Hanford, we must hasten the development of low cost atomic power. I think we should lead the world in this."

Twenty years later, nuclear electric energy in the Pacific Northwest had been transformed from a New Frontier vision posed by President Kennedy to an economic quagmire. Instead of being a model for the world in how to produce low-cost electric power, the WPPSS's nuclear energy program is in danger of being viewed as a monument to the financial futility of building nuclear power plants.

Traditionally, the Pacific Northwest has been served by two major segments of the electric utility industry—private investor-owned utilities and public power districts. Until recently, the region depended almost entirely on hydroelectric power. The investor-owned utilities built and operated generating facilities to meet their own needs. Some public power utilities also built and operated generating facilities, but most relied on the Federal hydroelectric power program, which evolved incident to construction of dams for navigation, irrigation, and flood control. The Pacific Northwest, particularly Washington State, has a long history of consumer-owned electric systems. But most of the present electric utility cooperatives came into being in 1930 after enactment of the Public Utility District (PUD) laws in Oregon and Washington. The PUD movement created a new dimension in public power. In fact, the PUD laws, together with the Federal hydroelectric program, created a new public power movement.

The Washington Public Power Supply System's entrance into the nuclear age was to a large degree influenced by a growing Federal role at the Hanford Nuclear Reservation near Richland, Washington. The Hanford works there, organized in April 1943 under the supervision of the Manhattan District of the Corps of Engineers, was built during World War II to produce plutonium used in nuclear bombs. In 1947, the Hanford project was placed under the jurisdiction of the newly created Atomic Energy Commission (AEC).

For some time there had been interest in multipurpose use of the Hanford facility. Some thought the waste heat from the federally owned Hanford reactors could be used to generate electricity. However, such proposals had been rejected, mainly because the Republican administration of President Dwight D. Eisenhower opposed Federal involvement in regional power development.

With the chill in U.S.-Soviet relations creating the "Cold War" in the 1950s, the AEC determined that additional military need for plutonium required that another production reactor be built at Hanford. In 1958, Congress authorized construction of a new reactor. Largely at the urging of U.S. Senator Henry Jackson of Washington State (and also, because of lobbying by the Bonneville Power Administration, the marketing agency for power generated by Federal hydroelectric projects in the Columbia River drainage basin) Congress and President Eisenhower agreed that the reactor would be "convertible," able to produce both plutonium and electricity. The authorization did not cover the necessary equipment, or even the building, for electricity production, however.

With the election of John F. Kennedy to the Oval Office, Eisenhower's restrictive policy regarding Federal involvement in regional power development came to an end. In the spring of 1961 Kennedy asked the Republican-dominated Congress to appropriate 60 million for an electric generation facility at Hanford to make use of the 13 million pounds of steam per hour that would otherwise be wasted. President Kennedy's request was strongly supported by the Department of Interior, the Bonneville Power Administration, the Atomic Energy Commission, the Federal Power Commission, and the American Public Power Association. But the House of Representatives, still controlled by people who had supported Eisenhower's restrictive policy voted it down.

Continued Congressional opposition to electric production at Hanford stemmed from the longstanding conflict between public and investor-owned power interests. The investor-owned utilities, supported by coal and railroad interests and the U.S. Chamber of Commerce, were simply too strong and too well organized for the advocates of public power. President Kennedy and Senator Jackson met defeat in the House of Representatives over the Hanford electric generation issue three times during the summer of 1961. It appeared that Federal funding for electrical generation facilities at Hanford was politically impossible.

As the House was voting against Federal involvement in the Hanford scheme, proponents continued to look for ways to achieve their goal. In anticipation of the House's third rejection, Washington State Governor Albert Rosellini instructed State Director of Conservation Earl Coe to investigate the possibility of Washington State's constructing and operating an electrical generating facility at Hanford. The idea was well received, and on the basis of initial discussions, State, and Federal entities reached a consensus at least on one point: The Washington Public Power Supply System would be the logical agency to undertake electrical power generation at Hanford.

WPPSS had been in existence about 4 years by that time. It was composed of 16 public utility districts (PUDs). As a joint operating agency of the State of Washington, WPPSS could build, buy, and operate electric generation and transmission facilities and sell electric power. It could also sell revenue bonds to finance its activities.

Organized in 1957, WPPSS had existed only on paper for several years. Then in 1960, the Lewis County PUD asked WPPSS if it would be interested in building the so-called Packwood Lake hydroelectric project. WPPSS' Board of Directors immediately seized the opportunity to get the agency into business, even though Packwood's electrical output would be small, only 27.5 megawatts. Bonneville Power Administrator Charles Luce went along with the plan by entering into a contract with the young agency to place all of Packwood Lake's electrical power output in BPA's Northwest distribution network.

WPPSS handled each step of the Packwood Lake project smoothly. PUD and State officials were pleased and impressed. With the Packwood Lake hydro project now completed and operating, WPPSS actively sought participation in the Hanford project.

Working with Conservation Director Coe was WPPSS' first managing director, Owen Hurd. Wasting no time, Hurd drafted a plan outlining how WPPSS could finance the Hanford project without Federal involvement and tie its 860 megawatts of power into the Bonneville Power Administration's grid. The plan made sense because WPPSS had successfully accomplished the same thing for the Packwood Lake hydroelectric project, and it was formalized on November 28, 1961.

Investor-owned utilities opposed the Hurd plan, but BPA Administrator Charles Luce liked it and began supporting it as BPA policy. Then the Atomic Energy Commission publicly endorsed the Hurd plan after learning that WPPSS would provide all the financing to modify the Hanford reactor to produce electricity, through the sale of a single issue of $130 million in revenue bonds. It was agreed in principle that the AEC would make available the waste steam from the reactor and that the Bonneville Power Administration would serve as electrical power distributor to WPPSS members.

BPA's and AEC's support of the Washington State moving in at Hanford caused furor immediately. In addition to outrage among investor-owned utilities, there was opposition from Congress' own General Accounting Office (GAO). On July 6, 1962, GAO stated that the Atomic Energy Commission did not have the authority to enter into a legal agreement with WPPSS and would need Congressional approval to do so. Once again WPPSS supporters people prepared for a difficult battle—but this time Senator Jackson offered a sweetner to the supporters of investor-owned utilities. He proposed to give 50 percent of the electrical output to investor-owned utilities if they would agree to the WPPSS/Hanford plan. Supporters of investor-owned utilities found the idea acceptable, and, although they did not publicly support the 50–50 split, they did drop their opposition and officially adopted a neutral position on the issue. The so-called compromise plan still failed to sway the pro-investor-owned House of Representatives, however, and on July 17, 1962, the House voted for a fourth time to prohibit the use of Hanford steam to produce electricity.

Following this fourth defeat in the House of Representatives, Senator Jackson, assisted by the supporting influence of the Kennedy White House, began one of the strongest lobbying efforts witnessed on Capitol Hill in many years. After two weeks of intensive lobbying, Jackson was able to get the Senate to agree to the WPPSS/Hanford plan. Following the Senate's approval, Jackson was also able to obtain the approval of the Joint Committee on Atomic Energy, of which he was a member, and finally of the House of Representatives. President Kennedy quickly signed the legislation containing the WPPSS/Hanford plan, and on September 26, 1962, he signed the Atomic Energy Commission Appropriations Act of 1963 (P.L. 87-701), which contained provisions for converting the Hanford reactor to electricity production.

With the WPPSS/Hanford plan now law, there was a flurry of activity. BPA, as promised, formally agreed to exchange electric power from its hydroelectric system for Hanford electric energy and worked out an exchange agreement under which a purchaser would obtain rights to BPA power at BPA rates. This arrangement included, in section 112(d) of the 1963 Act, the directive of Congress that:

> Any losses to the Bonneville Power Administration in connection with the arrangements or sales authorized herein shall be borne by its system customers through rate adjustments.

By January 14, 1963, 71 public agencies and investor-owned utilities had subscribed (in fact had oversubscribed) to Hanford's expected output. Half of the output went to the investor-owned utilities, as offered in the Jackson compromise plan, and the investor-owned utilities entered into the Hanford Exchange Agreements with BPA. On May 8, 1963, a $122 million bond issue (at 3.26 percent interest and $8 million less than previously estimated) was sold to finance construction of the WPPSS generating plant. Ground was broken on September 26, 1963, by President Kennedy.

Construction proceeded through 1964 and 1965, followed by testing in early 1966 and commercial operation on November 29, 1966. It was, for a while, the largest nuclear fuel generating plant in the United States.

Despite some early problems (frequent and lengthy outages), the Hanford facility generally has operated reliably, people in the Pacific Northwest felt that no other organization was in a better technical or organizational position to sponsor nuclear electric plants than the Washington Public Power Supply System. Nearly everyone missed one major point about the WPPSS/Hanford experience: WPPSS had not built the reactor or operated the reactor itself, but had only built and operated the generating portion of the facility, the part that turned the steam from the nuclear reactor into electric power. WPPSS had no experience in building or operating nuclear electric plants. The significance of this would become apparent later.

2

Need for Power

The Washington Public Power Supply System (WPPSS) nuclear plant construction program was based on projections of increased demand for power. Certainly, the region's population was growing. The 1970 census recorded population gains for each state in the Pacific Northwest, particularly Washington, which increased 19.5 percent over the 1960 population, to 3,409,169, and Oregon, which increased 18.2 percent, to 2,091,385. Industry in the region was increasing, too, thanks in part to long-standing Bonneville Power Administration policy to encourage industrial growth. BPA's policy was based on the widespread belief in the cause-and-effect relationship between energy and industrialization, between energy and jobs, energy and productivity, energy and gross national product, energy and the standard of living—indeed, between energy and the progress of civilization. BPA officials believed that low-cost power was the spigot of progress: when it was turned on, the economy flourished, and when it was turned off, the economy withered. Beginning with World War II and the imperative need for aluminum, BPA pushed into a long-term commitment to that industry.

In 1953, with the election of President Eisenhower, the industrial sales policy was taken away from BPA. Under-Secretary of the Interior Ralph Tudor ordered BPA to stop promoting new industrial power sales. The agency was to confine itself to selling surplus power and let investor-owned electric utilities in the Pacific Northwest supply all the uninterruptible power needed.

Certainly, BPA historic role was more narrow. It had been created in 1937 by the Bonneville Power Act, 16 U.S.C. §832, to market electricity generated at the Bonneville Dam on the Columbia River and to develop a transmission system for delivery of the electricity generated at this facility to customers throughout the Pacific Northwest. (Other relevant legislation is found in the Reclamation Act of 1937, as amended and supplemented, in the Flood Control Act of 1944, and the 1964 Northwest Power Preference

Act.) Over the years, BPA's authority was extended to include marketing the power from all Federal hydroelectric projects in the Columbia River drainage basin.

The Bonneville Power Act defined BPA's customers "preference" and "non-preference" customers. The Act required BPA to give preference and priority in marketing available power to public agencies and cooperatives. Specifically, the Act states:

> In order to insure that the facilities for the generation of electric energy at the Bonneville Project shall be operated for the benefit of the general public, and particularly of domestic and rural customers, the administrator shall at all times, in disposing of electric energy generated at said project, give preference and priority to public bodies and cooperatives.

Subject to the rights of preference customers, BPA was allowed to contract to sell excess power to private agencies and individuals.

With the election of President Kennedy in November 1960 and the appointment of Charles Luce as BPA Administrator in February 1961, BPA's industrial sales policy was reversed again. Luce, with the support and backing of the Kennedy Administration, adopted a positive planning policy aimed at an increased power supply and an aggressive power sales policy. One of the major elements of Luce's program was the Hanford electricity generating project. As a result of Luce's policy, the number of BPA contracts to supply power to industry surged. In BPA's 1961 annual report, Luce announced his intent to

> meet the load growth requirements of the region, including those for new industries whose location in the region is dependent upon the availability of low-cost power.

BPA's 1964 annual report devoted two pages to the "spectacular" increase in new industrial sales.

> This is the greatest period of industrial sales in BPA history, and commits nearly all the power BPA can sell to industry until new resources are assured.

Since the 1930s, the public utility districts and many other public utilities, energy cooperatives, and direct-service industries in the Pacific Northwest has purchased part, if not all, of their electric power from the Bonneville Power Administration. For several decades, the Pacific Northwest enjoyed an abundance of low-cost hydroelectric power.

Over the years BPA and its customers participated in a number of organizations that planned for the region's electrical needs. Beginning in the 1960s, the PUDs, other public and private power agencies, and BPA developed forecasts that available hydroelectricity would not be sufficient to meet all power needs in the future. Curiously, at the same time BPA was boasting about "booming" industrial sales, it was also predicting that it would soon run out of power. Why did BPA continue to push industrial sales when predictions as early as 1964 were that it would "run out" of power for industry in 1978 and for its preference customers (the public utility districts, for example) in 1982? The reasons were a combination of economics and a generalized belief in "progress." For example, in remarks to the Bonneville Regional Advisory Council meeting on December 5, 1966, BPA Power Manager Bernard Goldhammer itemized the benefits to BPA and the region from having a large amount of uninterruptible (high loading) power sales. The high loading of the Bonneville and Coulee Dam electric generators, he said, quickly put BPA on a sound business basis and also stimulated the earlier start of construction of a dozen hydroelectric projects. Industrial power sales helped keep BPA's costs low and thus helped ensure low power costs for the PUD preference customers and other utilities. High load factor industrial loads helped BPA ensure the stability and reliability of the transmission system and permitted the adoption of higher voltages and greater capacity. But mainly, Goldhammer saw economic benefits for the region:

> I feel that Bonneville's sale of power to industry has been good for the region, it has been good for our customers, it has meant lower wholesale power rates and lower retail rates than otherwise could have been achieved, it has helped diversify our economy, it has provided jobs, it has provided income for the area. And I think that the continuation of this policy will be good for the economy and for our customers.

Goldhammer's appeal notwithstanding, the Public Power Council (PPC), composed of the public and cooperative electric systems of the region, expressed concern about any more large blocks of power being sold to industry. Critics pointed out that aluminum plants used a lot of electric power but provided relatively few jobs.

When BPA was established in the 1930s, its founders believed BPA's series of hydroelectric dams on the Columbia River was like an oil well that would never run dry, a coal seam that would never thin out. Pacific Northwest political leaders and chambers of commerce referred to its unlimited potential.

By the late 1940s, utility leaders in the Pacific Northwest who recognized that the Columbia River does have limits were examining the region's coal resources. In the early 1950s, BPA's electric generators strained to meet the demand (loads) placed on them by the Korean conflict, and there was some Congressional interest in giving BPA limited authority to build and operate fuel-fired electric generating plants. The plan died in Congress, but interest remained, and in 1966, BPA was actively lobbying for permission to build thermal electric plants. BPA Administrator Luce's 1966 annual report (his last before leaving) suggested that thermal plants would be carrying much of the region's baseload by 1982. Incoming BPA Administrator David S. Black's transmittal letter in the same report said that the Pacific Northwest had reached the "pivotal point" in regard to thermal generation. The Washington State Joint Power Planning Council, chaired by BPA and included the region's utilities, started looking at the eventual construction of large nuclear and coal-fired electric generating plants.

Throughout the mid- and late 1960s, the use of nuclear power to supplement the limited hydroelectric potential left was vigorously explored by others in the Pacific Northwest. In a 1968 recommendation to the Washington State Legislature, the Legislature's Joint Committee on Nuclear Energy said:

> Nuclear power plants offer the most economical long-term solution to the State's power needs, and we must now turn to nuclear plants to supply our growing needs for industry, jobs, and an increasing standard of living. State government should be structured to participate in the timely and successful siting of nuclear power plants to balance our growing energy demands. Within state government, a procedure must be established wherein all state agencies having an interest in the siting of nuclear power plants may resolve their various interests into a single policy for the State of Washington and this enunciated

> formally as the policy of the state government. BPA should be encouraged to acquire power from nonfederal thermal plants.

The Joint Committee on Nuclear Energy was not the only governmental voice calling for nuclear power plants. Washington Governor Daniel Evans also looked to nuclear power:

> No market offers more promise than that of nuclear energy. We must be prepared to take the action necessary to assure maximum development of this great resource so that Washington becomes the nation's center of nuclear energy.

Now that a general consensus had been reached among the Pacific Northwest leaders and public officials as to the need for a long-term program of thermal energy development, the only remaining question was how to pay for it. Building the plants themselves, they assured each other, presented only minor technical difficulties. After all, hadn't they built some of the world's engineering marvels such as Hoover and Grand Coulee Dams? Financing a thermal construction program, on the other hand, constituted a formidable challenge.

Nearly all of the public utility districts in the Pacific Northwest depended on the Federal hydroelectric program operated by BPA for power—power they could purchase and then resell. As a result, most PUDs possessed virtually no collateral to offer in exchange for multimillion dollar loans through major financial institutions. BPA officials were keenly aware of this problem. In 1968, BPA Administrator H.R. Richmond requested that the Washington State Joint Power Planning Council (JPPC), an organization composing of BPA and 110 public and investor-owned electric utilities, study the possibility of adding a number of large thermal-electric generation plants to the Pacific Northwest power grid.

After considerable study, in October 1968, the JPPC agreed on a $15 billion, 20-year program to construct new thermal, hydro, and transmission facilities (see Table 1). By January 1969, the estimated cost of the Hydro-Thermal Power Program (HTPP), as it was called, had increased to $17.9 billion for the period through July 1990. Of this amount, $6.1 billion was to come from Federal funds to build BPA's transmission facilities.

The HTPP, as formulated in 1968, was a planning document, a declaration of intentions. It showed on paper how the 20-year projected demand (load) growth could be served by building new

Table 1
Joint Power Planning Council's Hydro-Thermal Power Program, 20-Year Construction Plan, 1000 Megawatt Projects

Project	*Operation Date*	*Ownership*
*Centralia #1	Sept 1971	Joint
*Centralia #2	Sept 1972	Joint
Nuclear #1	Feb 1975	Investor
Nuclear #2	May 1975	Investor
Nuclear #3	April 1977	Public
Nuclear #4	Oct 1978	Investor
Nuclear #5	April 1980	Investor
Nuclear #6	Jan 1982	Investor
Nuclear #7	March 1982	Public
Nuclear #8	Oct 1983	Investor
Nuclear #9	Feb 1984	Public
Nuclear #10	April 1985	Investor
Nuclear #11	Jan 1986	Investor
Nuclear #12	March 1986	Public
Nuclear #13	Jan 1987	Investor
Nuclear #14	April 1987	Public
Nuclear #15	Jan 1988	Investor
Nuclear #16	May 1988	Public
Nuclear #17	Jan 1989	Investor
Nuclear #18	March 1989	Public
Nuclear #19	Nov 1989	Investor
Nuclear #20	Feb 1990	Public

*These two are 700 megawatts.
Source: BPA, *Hydro-Thermal Power Program* (1969).

---

generating facilities. During 1969 the JPPC refined the more general 20-year program into a 10-year program to meet the region's needs through 1981. The ten-year program, known as HTPP, Phase 1, called for a gradual transition from the use of only hydroelectric power for baseload energy to a mixed base of hydroelectric and thermal generating resources, with any new hydroelectric resources to be used to increase peaking capacity. The controlling concept was the planning, construction, and operation of the region's power facilities as if they were under a single ownership. The utilities, both investor-owned and public, would build the thermal plants. BPA would provide the necessary peaking capacity, the high-voltage transmission grid, and reserve capacity. The immediate goals for HTPP, Phase 1 included the expansion of BPA's transmission system, the addition of new hydroelectric generators for peaking, and the construction of three coal-fired and four nuclear plants, including WNP 1, 2, and 3. Six of these seven plants would be paid for through "net-billing" contracts with BPA.

The "net-billing" concept was devised because BPA lacked statutory authority to build or own any interest in non-hydroelectric generating plants. For each project, WPPSS, BPA, and the several so-called "Participants" would enter into "Net-Billing Agreements." Under these agreements, WPPSS would own and operate the nuclear projects. Each "Participant" would buy a specified percentage of each project's generating capacity and would be obligated to pay the same percentage of the project's annual costs. In fact, the Participant would assign its share of the project's annual costs, whether or not the plant was operating or operable. To make these payments, BPA would credit the Participant's bill for power received from BPA in the amount the Participant paid to WPPSS or, in the event of a shortfall, would pay cash, either to the Participant or directly to WPPSS. Put simply, under the "Net-Billing Agreements" BPA purchased power and paid for it by adjusting the amount owed to BPA for power previously delivered to the participating preference customers under their BPA Power Sales Contracts.

The "net-billing" scheme effectively bonded BPA, the public utilities, the PUDs, and WPPSS together in one power-plant planning, financing, and operating family. BPA would in effect provide the financial security, the collateral, that would allow the PUDs and the others such as WPPSS to sell revenue bonds to finance the projects. Indirectly, BPA would be backing the thermal power plants outlined in HTPP, Phase 1 with the financial resources of the U.S. Government. The PUDs would construct and operate the power plants but would transfer all the electrical output to BPA. BPA would place the energy in its Pacific Northwest distribution grid network, and the PUDs would collect money from their customers, in the form of changes, to pay interest and principal on the revenue bonds. Money left over after the PUDs had paid their bills would be forwarded to BPA so that the costs between the thermal energy and the hydro energy could be melted together to help hold down electricity rates.

Members of the Joint Power Planning Council were pleased with themselves, for they had hammered out a pretty neat arrangement—an energy plan with something in it for everyone. Electric energy demand throughout the Pacific Northwest could be met well into the next century, all new thermal plants would be under investor-owned and public utility ownership, and BPA would be able to satisfy all of its power customers and carry out its historic goals of regional economic development. The legal and financing problems were effectively circumvented for everyone, and the whole deal was capped off with the billing arrangement between the PUDs and BPA.

BPA's net billing arrangement with Pacific Northwest PUDs was approved by President Richard M. Nixon on October 27, 1969. Similar programs in other regions of the United States were encouraged, and Secretary of the Interior Walter Hickel modified the Interior Power Policy on September 30, 1970 to read:

> The Department encourages the construction and operation of electric power facilities by non-Federal interests that will develop and protect the natural resources in full accord with the broad public interest. All segments of the electric power industry are encouraged by the Department to cooperate in the planning, design, development, and pooling of power generation and transmission facilities on a regional, interregional, and national basis which will provide an adequate and reliable supply of electricity.

In the Public Works Appropriation Act of December 11, 1969, Congress approved the net-billing arrangement with respect to the Trojan nuclear plant to be built at St. Helens, Oregon. Later, on October 7, 1970, the Public Works Appropriation Act was broadened so that the net-billing arrangement could be applied at the other five proposed power projects outlined in Phase 1 of the Hydro-Thermal Power Program.

3

Bonneville Power Administration and the Washington State Public Utility Districts

In Bonneville Power Administration's estimation, the need for new non-hydroelectric power was unquestioned. So great was the need, in fact, that BPA and the region's energy forecasters in 1968 projected building 20 nuclear reactors in the Pacific Northwest grid by 1990.

Ken Billington, former Executive Director of the Washington Public Utility District Association, remembers the first expressions of concern about meeting projected demand. The year was 1956 and President Dwight Eisenhower was talking about selling off the Federal government's Columbia River dams. Many of the State's public utility districts, which depended on power generated by those Federal dams and sold through the Federal Bonneville Power Administration, were nervous. For four years the PUDs had been parrying with the State's Republican administration over their desire to form a joint agency to build power plants. "Our loads were jumping," recalls Billington. "There was talk of no more dams on the Columbia. It meant local PUDs would have to get together."

In 1953 the PUDs' wishes had been partially granted when the Washington State Legislature approved a bill that would allow them to form a joint operating agency. In 1957, 16 public utility districts joined together to form a joint operating agency which they called the Washington Public Power Supply System. Later WPPSS added 3 more PUDs, bringing the total to 19.

"We [public utility districts] were generally the 'have nots,' not 'the haves,'" recalls Glenn Walkley, WPPSS Board president for 10 years and a six-term Franklin County PUD Commissioner. "[We] didn't own generating plants or dams like the cities of Seattle or Tacoma. All of us were too small or too inexperienced to try building generating facilities alone." WPPSS would make it possible for those utilities to band together to buy up the Federal government's dams, if they were ever put on the auction block (they never were), or to build dams or power plants of their own.

WPPSS' first project, the small hydroelectric dam at Packwood Lake in Lewis County, Washington, recalls Billington, "was what WPPSS used to prove up its ability to raise money and to successfully build and operate a power plant." WPPSS's next project, the generating plant constructed to take advantage of steam produced at the Hanford Nuclear Reservation, "was considered a coup for us PUDs over the IOUs (investor-owned utilities)," says Walkley. The WPPSS/Hanford plant was the largest nuclear electric generating plant in the country at that time.

In the late 1960s, BPA, the aluminum industry, and the Public Power Council began looking at the need for more power in the decades to come. "Our loads were jumping 6, 7, to 10 percent a year," Billington recalls, and the forecasts at the time predicted the need for thousands more megawatts of power. BPA began telling all its customers that it would be unable to provide electricity to match load growth in their service areas after 1983. It told the aluminum industry it would stop serving it directly when its power contracts expired. In the meantime, the PUDs had been stymied in plans to build new hydroelectric dams on the Snake and the proposed Ben Franklin Dam on the Columbia River. Said Walkley: "BPA told us it would be cheaper to build nuclear plants than coal plants, so we took their advice."

BPA continued in the late 1960s to early 1970s to predict shortages of electric energy in the Pacific Northwest and to advise preference customers that they would be responsible for developing new generating resources to serve their load growth. For example, on September 14, 1972, Donald Hodel, then BPA Deputy Administrator, in describing the region's power supply future, said to the Electric Marketing Conference of the Northwest Public Power Association:

> Perhaps some of you are sitting there wondering why I dwell on this problem. You may be thinking that "we don't have to worry about it. We are protected by the preference clause and we will be able to go on buying power from BPA. BPA will have to withdraw its service to investor-owned utilities and then to industrial customers in order to sell to us." Consider this: In a time of regional shortages the preference clause may not mean very

> much. For then it may come to a political decision as to who gets the power.

Frequently BPA publications promoted the need for nuclear power with almost missionary zeal:

> Failure to meet regional power requirements as they develop will result in economic and social penalties. Increased use of electricity has contributed importantly to the emancipation of peoples from poverty and drudgery and to expansions in human capacity to live the good life. Any further efforts to extend these benefits to a larger segment of our society will require more electricity. Significant growth and power requirements cannot be stopped without compromising the future well-being of people. Failure to meet future power requirements is an unacceptable alternative.

During the mid- to late 1960s and early 1970s BPA, its own forecasts of energy shortages notwithstanding, offered its direct service industrial (DSI) customers modified 20-year uninterruptible power sales contracts. The effect of these contracts was to extend the duration and the size of BPA's obligation to deliver power.

By the early 1970s, when the decisions were being made on WPPSS' role in the nuclear building program, both Seattle and Tacoma municipal light companies joined WPPSS. "Though some of those utilities may now rue their decision to join WPPSS' nuclear projects, they had perfectly good reasons for participating at the time," says Billington. "Put yourself in their shoes. Their power demands were jumping 4 to 6 percent a year, too. BPA told them it would have no more power and some had the prospect of having to serve a large aluminum company they had not served before. I think most of us would have made the same decisions."

In September 1973, WPPSS issued municipal tax-exempt bonds and construction began on WNP 1, 2, and 3. Because of the "Net-Billing Agreements," BPA was obligated to pay for these costs of construction. Ultimately, of course, these costs would be passed through to BPA's customers in the form of higher rates for energy sold under the customers' Power Sales Contracts.

The Origins of WNP 4 and 5

By early 1973, regional planners had concluded that the Hydro-Thermal Power Program would have to be modified to provide for the region's power needs beyond 1981. Planned thermal plants had experienced delays, demand had grown faster than had been forecast, and the capacity for net billing was becoming exhausted.

After extensive consultation, BPA and its customers agreed on December 14, 1973, to a cooperative plan titled "Phase 2 of the HTPP." Under Phase 2, the utilities were to construct eight additional plants with a total capacity of 7.56 million kilowatts (megawatts) to meet the region's baseload power needs through 1986 (additional plants were to be added later to meet requirements through 1994). The Federal government (through BPA) was to construct facilities for additional peaking capacity, totaling approximately 3.7 million kilowatts and also was to expand its transmission system.

Phase 2 of the HTPP differed from Phase 1 in that BPA could not use net-billing agreements to acquire the generating capacity of additional thermal generating plants constructed by its preference customers. This was so for two reasons: (1) there was insufficient net-billing capacity, and (2) a recent U.S. Department of the Treasury regulation promugated under Section 103(b) of the Internal Revenue Service code now prohibited the assignment of power from new plants financed by tax-exempt municipal bonds to non-exempt purchasers, which included BPA. Preference customers would have to meet increased baseload energy requirements beyond 1983 by particpating directly or indirectly in the construction of new plants. As part of Phase 2, WPPSS, in May 1974, commenced planning for the financing and construction of two more nuclear plants, WNP 4, to be located adjacent to WNP 1 at Hanford, Washington, and WNP 5, to be located adjacent to WNP 3 at Satsop, Washington.

No "Dry-Hole" Risk to Public Agency Sponsors

An integral part of HTPP, Phase 2 as proposed to all Participants and other BPA preference customers in December 1973 was that the entire region, not just the sponsoring utilities, would bear the costs of a "dry hole" (that is, customers throughout the region would have to pay to-date costs even if no energy was ever produced). In a letter to all PUD commissioners and managers dated

December 18, 1973, Ken Billington, Executive Director of the Washington PUD Association, explained:

> A means to provide regional underwriting of a "dry hole" generating plant was proposed. This would be accomplished by using BPA as a contractual "centering post" whereby the sponsors of a generating plant which had been accepted as a regional resource would be reimbursed for losses sustained because such plant became inoperable for reasons beyond prudent utility management. Those funds would come from an assessment made by BPA against every kilowatt hour of electricity sold to an ultimate customer by every utility and BPA in the Pacific Northwest. Definition will have to be made as it is not intended that the term "dry hole" would cover plants delayed in construction or which do not come up to planned and forecasted plant operational levels. It was agreed that no one utility or group of utilities should have to bear losses on a plant being built for a regional purpose which does not materialize and which results from factors beyond the control of the involved utility or utilities.

In a "program summary" of HTPP, Phase 2, by Alan H. Jones, Chairman of the Public Power Council (PPC), on December 20, 1973, the following explanation was given to all PPC members:

> "Dry-Hole"—It was agreed that no one utility or group of utilities should have to bear losses on a plant built for a regional purpose which does not materialize and which results from factors beyond the control of the involved utility or utilities.

On January 25, 1974, BPA's Power Manager, Bernard Goldhammer, sent to all preference customers a summary of HTPP, Phase 2 and a cover letter which stated that the entire region would share "dry-hole" costs. At the October 9, 1975, meeting of PUD managers, Ken Billington mentioned a concept for an "BPA Backup" for HTPP, Phase 2, including a guarantee in case of possible default by the Participants in case of a "dry hole."

In August 1974, Larry Hittle, a BPA lawyer, distributed and explained to a PPC committee a preliminary draft of a "Trust Agency Agreement" that would prevent any losses to PUDs in case of a "dry hole." On December 11, 1974, at the annual meeting of Washington State PUD managers, the PUD managers were assured by BPA spokesmen, that "there is a dry-hole coverage in the [Trust

Agency] Agreement." Although BPA continued to represent through 1975 to its preference customers that it would enter into a Trust Agency Agreement with those PUDs that desired it, no Trust Agency Agreement was ever formally offered by BPA.

New BPA/DSI Power Contract

At the time BPA was talking about a Trust Agency Agreement to prevent any losses to PUDs in case of a dry hole and it was predicting that it would not have sufficient power to meet the projected demands of its preference customers and its contractual obligations to non-preference purchasers, BPA was offering new noninterruptible power contracts to its direct service industrial (DSI) purchasers. These industrial contracts extended the terms and increased the size of BPA's obligation to deliver power to nonpreference customers. BPA knew that by entering into these new industrial contracts, it was jeopardizing its ability to serve the full contractual needs of its preference customers.

Litigation Stops BPA's Participation in HTPP, Phase 2

On August 27, 1975, the Federal District Court in Oregon held that BPA could not participate in developing new electricity generating plants until it filed a comprehensive environmental impact statement on its activities. Earlier in the year, on April 2, 1975, BPA had submitted a draft environmental impact statement. However, the court viewed HTPP, Phase 2 as a major Federal action requiring a comprehensive EIS. The ruling required that BPA explicitly define its role in the Pacific Northwest power supply system. Furthermore, the court ruled that BPA's power supply contracts with the DSIs could not be enforced until BPA prepared the EIS. From this decision, BPA determined that it could not undertake any acts in furtherance of HTPP, Phase 2 until a comprehensive EIS had been prepared and approved. This meant that, for the time at least, BPA would not be providing any financial guarantees for additional power plants. More important, perhaps, BPA would not be assuming any kind of leadership in planning the HTPP program. It could continue to closely monitor and, when possible, indirectly aid the power developers, but for now it had to withdraw from HTPP power planning, construction, and operation activities.

Notice of Insufficiency

Under the general provisions of its Power Sales Contracts with its customers, BPA had the right to restrict power deliveries to preference customers should Federal power supplies be inadequate to meet the needs of those customers. If it wanted to exercise this right, BPA had to give prior written "Notice of Insufficiency" to its customers. Under various amendments to the Power Sales Contracts, the "Notice of Insufficiency" had to be issued five to eight years prior to restricting delivery.

As early as 1972, BPA had noted both in public forums and in written documents that if preference customers did not participate in HTPP, Phase 2, it might have to issue Notices of Insufficiency in 1973. In 1973, U.S. Secretary of the Interior Rogers Morton instructed BPA Administrator Donald Hodel to informally notify BPA preference customers that BPA would be forced to issue Notices of Insufficiency unless they agreed to a plan to ensure the region's future electric power supply.

During the period 1972 through May 1976, BPA reiterated on numerous occasions its intention to issue Notices of Insufficiency to preference customers, should HTPP, Phase 2 not be implemented. On May 5, 1975, BPA Administrator Hodel stated in writing that if the preference customers did not sign all the contracts related to Phase 2 by July 1, 1975, BPA would issue Notices of Insufficiency. Customers did sign interim Option Agreements, and BPA delayed issuing Notices. However, BPA continued to suggest that it would issue Notices, and in April 1976 it did circulate to its preference customers a draft "Notice of Insufficiency."

BPA's Promotion of HTPP, Phase 2

As it explored ways to implement HTPP, Phase 2, BPA determined that WNP 4 and 5 should be sponsored by the region's preference customers—that is, by public utilities—so that the projects could be supported by tax-exempt financing. Thereupon, BPA embarked on a region-wide program to promote sponsorship of WNP 4 and 5 by its preference customers.

From 1973 through 1976, BPA representatives conducted a series of meetings throughout the Pacific Northwest at which they informed BPA preference customers of the need for WNP 4 and 5 and encouraged them to sponsor the two projects, warning that

> if there was a power shortage in the region, the preference clause might not mean very much, and the allocation of available Federal power might be a "political decision."

To further encourage its preference customers to agree to sponsor WNP 4 and 5, and to allay fears about potential financial risks, BPA represented to the preference customers that it would seek statutory authority to acquire the output of WNP 4 and 5, thereby removing the financial risk to the sponsors. As a result of these statements, some of the participants may have concluded that BPA would eventually assume some kind of financial responsibility for WNP 4 and 5. Former BPA Deputy Administrator Ray Foleen concedes that "BPA certainly made a case for signing [the agreements]." However, Foleen continues, "there was nothing at this time that promised that BPA would assume financial responsibility of the plants." In letters to potential participants, BPA Administrator Donald Hodel made it clear that BPA did not have any existing plans to join in WNP 4 and 5's construction financing.

To implement preference customer sponsorship of WNP 4 and 5, BPA was involved, along with WPPSS, in drafting a series of contracts, or agreements, to be signed by participants and helped distribute these documents to preference customers. These documents were to be executed by WPPSS for each participating utility.

Preliminary Planning for WNP 4 and 5

In May 1974, WPPSS began planning for the financing and eventual construction of WNP 4 and 5. Late in 1974, WPPSS issued $17.5 million of revenue notes, to cover preliminary planning on the strength of proposed (but unsigned) agreements between BPA and its preference customers. On January 1975, BPA wrote to "All Bonneville Preference Customers" about the status of the various agreements necessary to implement HTPP, Phase 2. Noting that preliminary planning funds would soon be exhausted and that additional financing was needed, BPA stated:

> If additional money is to be obtained, agreements should be executed by all parties by early May. This requires that our negotiations regarding the Trust Agency Agreement, the Proposed Power Sales Agreement, the WPPSS Participant Agreement, and

United States Department of the Interior

BONNEVILLE POWER ADMINISTRATION

P.O. BOX 3621, PORTLAND, OREGON 97208

OFFICE OF THE ADMINISTRATOR

eply refer to: P

JUN 28 1976

Mr. William Hulbert, Manager
Snohomish County PUD No. 1
P.O. Box 1107
Everett, Washington 98206

Subject: Contract No. 14-03-29131 (Power Sales Contract) - "Notice of Insufficiency"

Dear Mr. Hulbert:

The Bonneville Power Administration has the obligation under the Power Sales Contract to meet the District's power requirements, subject to limitations included in such contract, unless, pursuant to Section 22 of the General Contract Provisions attached to such contract, Bonneville determines that Bonneville's resources will not be adequate to meet its estimated loads, subject to specific notice requirements.

Bonneville has completed an analysis of the resources it estimates will be available for disposition and its requirements and commitments to supply firm energy in the year July 1, 1983, to June 30, 1984. As a result of this analysis, Bonneville has determined that the firm energy resources available to it will be insufficient in that year to supply in full the District's firm energy requirements, the firm energy requirements of other preference customers, and Bonneville's obligations to deliver firm energy to its other customers.

Therefore, in accordance with the provisions of the Power Sales Contract, I hereby give notice, effective at 2400 hours on June 30, 1976, that in the year beginning July 1, 1983, and in each year thereafter during the term of the Power Sales Contract, Bonneville's obligation to supply firm energy to the District will be limited to an allocation, the amount of which will be computed according to the terms of Section 22 of the General Contract Provisions.

NOTED
JUN 28 1976
W. G. HULBERT JR.

Letter to William Hulbert, Snohomish County PUD, Subject: Notice of Insufficiency - Contract No. 14-03-29131

Bonneville has provided the District and its other customers with a preliminary forecast dated April 6, 1976, of the District's allocation based upon then available information. Bonneville will furnish the District with a new forecast when data is available regarding the District's 1975-1976 net system firm energy load. From time to time as additional data becomes available, Bonneville will furnish the District with revised forecasts of its allocation and a final notice of allocation in accordance with such Section 22.

If you have any questions concerning this letter, the disposition of power after the date of insufficiency, or your allocation, please contact your Bonneville Area or District Office.

Sincerely yours,

Donald Paul Hodel

Administrator

> the WPPSS Industrial Sales Agreement be in final form by March 1, 1975. All parties are optimistic that this can be achieved.
>
> However, the Public Power Council, Bonneville, and WPPSS have recognized that mere optimism on the conclusion of these contractual negotiations is not enough. We are preparing a proposed interim arrangement which could be executed by all 110 of Bonneville's preference customers to provide backing for any additional financing WPPSS will need to keep the projects on schedule.

Many of the participating customers failed to respond to BPA's call for a quick wrap-up of the contractual arrangements. The reason was simple enough. The financing of WNP 4 and 5 would not be backed by BPA, as was WNP 1, 2, and 3. The participating utilities would be directly financing WNP 4 and 5 through their own borrowing capacities. Naturally, investors in the bond market would view bonds for WNP 4 and 5 as having greater risk than bonds for WNP 1, 2, and 3 because they lacked Federal backing. To secure bond sales, investors would have to be motivated by higher interest rates, thus adding to construction costs.

Because of the continuing hesitancy of preference customers to participate in WNP 4 and 5, BPA, in a letter dated February 25, 1975, proposed an interim solution in the form of an Option Agreement. Under the Option Agreement, preference customers would secure additional development funds for WNP 4 and 5 by issuing $100 million worth of tax-exempt municipal bonds, which WPPSS called "development bonds." BPA mailed the proposed Option Agreement to its preference customers on March 3, 1975. By July 1975, 93 preference customers had expressed interest in signing Option Agreements with WPPSS, and the development bonds were sold on July 24, 1975. (Only 88 of the 93 preference customers would actually sign the Option Agreements.) Thus, initial development and construction of WNP 4 and 5 was commenced by WPPSS under these Option Agreements.

After the development bonds were sold, an essential element of the WNP 4 and 5 financing scheme, Assigned Agreements between the participants and WPPSS, were prepared. Under these

Opposite: Bonneville Power Administration's June 1976 "Notice of Insufficiency" Letter.

Table 2

Participants in WNP 4 and 5 Arranged by State

(In parentheses: M = Municipal, P = Public Utiliy District, C = Co-op)

Washington	*Share*
Alder (M)	.00011
Benton (P)	.05080
Benton Rural (C)	.00670
Big Bend (C)	.00514
Blaine (M)	.00067
Centralia (M)	.00659
Chelan (P)	.00642
Challam (P)	.01373
Clark (P)	.09858
Columbia (C)	.00647
Cowlitz (P)	.09132
Douglas (P)	.00011
Ellensburg (M)	.00625
Elmhurst (C)	.00581
Franklin (P)	.02925
Grant (P)	.00581
Grays Harbor (P)	.04410
Inland (C)	.02244
Klickitat (P)	.00982
Lewis (P)	.02021
Lincoln (C)	.00190
Mason #1 (P)	.00156
Mason #3 (P)	.00971
McCleary (M)	.00123
Nespelem (C)	.00045
Ohop (C)	.00089
Okanogan (C)	.00045
Okanogan (P)	.00681
Orcas (C)	.00647
Pacific (P)	.00848
Parkland (P)	.00134
Pend Oreille (P)	.00402
Port Angeles (M)	.00469
Richland (M)	.01965
Skamania (P)	.00257
Snohomish (P)	.13051
Steilacoom (M)	.00145
Sumas (M)	.00022
Tacoma (M)	.10696
Tanner (C)	.00100
Vera (Irreg.)	.00257
Wahkiakum (P)	.00123

Oregon	*Share*
Bandon (M)	.00067
Blachly-Lane (C)	.00458
Canby (M)	.00525
Cascade Locks (M)	.00067
Central (C)	.00971
Central Lincoln (P)	.02668
Clatskanie (P)	.00781
Columbia Basin (C)	.00391
Consumers Power (C)	.01351
Coos-Curry (C)	.00581
Douglas (C)	.00514
Drain (M)	.00067
Hood River (C)	.00301
Lane (C)	.00770
McMinnville (M)	.00971
Midstate (C)	.00703
Milton Freewater (M)	.00056
Northern Wasco (P)	.00324
Salem (C)	.00458
Springfield (M)	.01764
Tillamook (P)	.00781
Umatilla (C)	.03573
Wasco (C)	.00134
West Oregon (C)	.00134

Idaho	
Bonners Ferry (M)	.00190
Burlev (M)	.00190
Clearwater (C)	.00324
Fall River (C)	.00648
Heyburn (M)	.00257
Idaho County (C)	.00045
Idaho Falls (M)	.00915
Kootenai (C)	.00647
Lost River (C)	.00134
Northern Lights (C)	.00514
Prairie (C)	.00089
Raft River (C)	.00391
Rupert (M)	.00324
Rural (C)	.00089
Salmon River (C)	.00084
Unity (C)	.00134

Wyoming		*Montana*	
Lower Valley (C)	.00837	Glacier (C)	.00179
		Missoula (C)	.00581
Nevada		Ravalli (C)	.00234
Wells (C)	.00045	Vigilante (C)	.00290

Total Washington State .74449 percent or $1,674 billion.
Total non-Washington State .25547 percent or $576 million.

---

agreements, WPPSS was assigned the output of WNP 4 and 5 for the purpose of selling the power to direct service industries (DSIs). Eighty-seven of the 88 Participants in WNP 4 and 5 executed Assigned Agreements with WPPSS. Another agreement was drafted —the "Short Term Sales Agreement"—which enabled WPPSS to sell the output of WNP 4 and 5 to the DSIs and obligated the DSIs to purchase a certain amount of the power so offered. However, neither agreement obligated the DSIs to anything if WNP 4 and 5 never became operational. As a result, the financial risk of a "dry hole" remained with the Participants. In fact, none of the agreements that ultimately implemented the WNP 4 and 5 program provided "dry hole" protection for the Participants.

By April 1976, WPPSS still had not signed Participants' Agreements with all 88 Participants. This worried WPPSS and BPA because Washington State law required joint operating agencies to obtain full participation agreements on a plant before they issued revenue bonds. So BPA met again with its preference customers and reaffirmed that it could not meet their future power requirements and soon would issue Notices of Insufficiency. Then on June 24, 1976, while most of its preference customers were considering whether to sign the Participants' Agreements, BPA officially issued Notices of Insufficiency to its preference customers. This had a profound impact on BPA's full-requirement customers, the PUDs that were totally dependent on BPA for their electric power needs. Facing a possible electric energy shortage, and having no feasible alternative, the PUDs had no reasonable choice other than to participate in WNP 4 and 5.

By June 24, 1976, all 88 Participants had finally entered into Participants' Agreements with WPPSS. By doing so, each Participant purchased a percentage share of the generating capacity of WNP 4 and 5, which were to be constructed and operated by WPPSS (see Table 2). The Participants could use, sell, or otherwise dispose of this capacity. In exchange, each Participant promised to pay WPPSS a portion of the annual budget for constructing and operating WNP 4 and 5, according to its percentage share. In effect, the Participants assumed obligations for WNP 4 and 5 similar to those BPA had assumed for WNP 1, 2, and 3 under the Net-Billing Agreements. After all 88 Participants' Agreements were signed, a resolution allowing the sale of project bonds was passed by the Board of Directors, bonds were issued by WPPSS, and the construction of WNP 4 and 5 was begun. Bonds in the face value amount of $2.25 billion were issued; with interest, the total debt amounted to approximately $7.0 billion.

Subsequent to the execution of the Participants' Agreement and the commencement of construction of WNP 4 and 5, a number of regional electric energy load forecasts that questioned BPA's need for the power to be produced by WNP 4 and 5 were issued. However, it was too late for any more discussion. All 88 Participants' Agreements had been signed and returned to BPA. The Participants had agreed to join in the building of WNP 4 and 5, even in the absence of any financial security should the two nuclear projects not be built.

4

Seattle Residents Order Their City Light Company Not to Participate in WNP 4 and 5

Had more people asked questions like those asked in Seattle regarding municipally owned Seattle City Light's proposed participation in WNP 4 and 5, it is unlikely that WPPSS and its Participants would have economic woes of the magnitude they now face. In the early 1970s City Light, with the approval of its "boss," the Seattle City Council, purchased a 10 percent share of WNP 1 without much fanfare. In 1974, City Light officials, believing they needed even more electric power for the late 1980s and 1990s, wanted to purchase a 10 percent share of WNP 4 and 5. The City Council had accepted City Light's argument about the need for power, had approved participation in WNP 4 and 5, and was virtually ready to hand over $10 million to WPPSS as the first partial payment for its 10 percent share (later, City Light would have to contribute up to $250 million for its full 10 percent share). First, however, City Light needed to prepare a short statement about the likely environmental impact on Seattle of building WNP 4 and 5, as required by the Washington State Environmental Policy Act. The payment would be delayed for only a few days while someone at City Light got around to writing the statement. Then Gordon Vickery, Manager of City Light, got the idea of having a local citizen's group prepare the environmental impact statement—and before it was all over, the Seattle City Council had pulled out of WNP 4 and 5. The Council decided that the City of Seattle would have all the electrical energy it needed for the 1980s and 1990s by undertaking a massive energy conservation program.

How Seattle Citizens Said No to WPPSS

In 1973–1974, Seattle City Light participated with other Pacific Northwest electric utilities, in the planning of WNP 4 and 5. When the time came for signing up for WNP 4 and 5 during the summer of 1974, the City Council, with virtually no questions asked by Council members or the public, approved the 10 percent buy-in.

About the same time, Donna Klemka of the Seattle branch of the Sierra Club, along with representatives of other environmental groups, started tracking nuclear power plant construction and operation and the impact on the Pacific Northwest's environment. In the spring of 1974, when City Light first mentioned the need for money to sign up for WNP 4 and 5, Klemka began wondering how City Light was going to handle the requirements of the Washington State Environmental Policy Act (SEPA). She asked Peter Henault, an environmentalist for City Light whom she knew and who shared her interest in environmental matters, how City Light was going to comply with the SEPA regulation. Henault didn't know, but he asked his fellow officials. Their response was that City Light had participated in WNP 1 without any environmental questions being asked. Henault noted that times had changed and said that City Light would have to comply. He recommended that the utility file a short environmental report, as called for in SEPA regulations, but his recommendation was brushed aside. City Light officials told him he was being too much of a purist and that his worry over SEPA was unnecessary. There was no indication that the general public in Seattle was the least bit concerned over City Light's participation in WNP 4 and 5, they said, and there was no indication that SEPA people were concerned, either. Henault suggested that City Light at least submit a declaration of "nonsignificant impacts," but officials told Henault that even this was unnecessary.

Informed of City Light's decision to ignore the SEPA requirement, Klemka wrote to environmental groups in Washington State saying, "You utility is about to make a financial commitment to participate in the ownership of several nuclear electric power plants to be constructed by WPPSS." "In addition," she wrote, "City Light has no intention of providing the required environmental background information required by SEPA." She called on her colleagues to write to the Seattle City Council asking how it intended to comply with SEPA.

When the Seattle City Council received one of these inquiries, it automatically forwarded it to Gordon Vickery, Manager of City Light, with the notation, "Take care of this." With the receipt of the letter, Vickery and others at City Light became worried. Vickery wanted Henault to write a response saying that there wasn't time to comply with SEPA regulations. Henault said no, that in good conscience he couldn't do that. Again he suggested that a declaration of nonsignificant impacts be written. Vickery repeated what he had said earlier, that such a declaration wasn't necessary. Instead, he said, "We'll take the position that prior environmental impact statements written by WPPSS on WNP 4 and 5 for NEPA [National Environmental Policy Act] have already satisfied the requirements of SEPA." This was the position taken by City Light. However, the Washington Environmental Quality Council (WEQC) didn't agree with City Light's assertion, and Vickery found that the SEPA issue was getting out of hand.

City Light's decision to ignore SEPA caused a public outcry. Environmental groups demanded that the City Council hold public hearings on the environmental impact of City Light's 10 percent share of WNP 4 and 5. Because of the growing controversy, the City Council requested that spokespersons from Bonneville Power Administration and the U.S. Department of Energy's (DOE) Seattle office appear before it and provide some background information on WNP 4 and 5. The City Council was told by these energy experts that Phase 2 of the Hydro-Thermal Power Program was good and that City Light needed to have its 10 percent share of WNP 4 and 5.

At this time, area environmental groups felt they did not have enough information to continue to press their case that WNP 4 and 5 might be unnecessary. In addition, several members of the City Council weren't sure, one way or the other, of what they had heard from BPA, DOE, and their own City Light. They wanted more information. They agreed to give City Light $10 million for preliminary spending on WNP 4 and 5, but they told City Light officials that

> Before you return in another year [requesting] full participation in these two projects and the $250 million, we'll want you to answer the questions the environmental groups, the general public, and our own Council members have raised regarding our participation in WNP 4 and 5.

Gordon Vickery asked his environmental manager Peter Henault to direct the investigation. Vickery liked the plan Henault outlined and immediately approved it. Henault had recommended something that neither City Light nor the City Council had requested—some form of citizen participation—but that didn't bother Vickery. He had long been an advocate of citizen involvement in government decision-making. When he was Seattle's fire chief, he had set up a number of citizen involvement committees to help select fire station houses, a program that had been highly effective and popular with the public. Even Henault's selection to the post had come through citizen involvement. In 1972, when Vickery became manager at City Light, one of his first actions was to form a citizens committee composed of environmentalists and others to select a person for the environmental manager's job. Henault had been the committee's choice.

Before starting public involvement activities Henault wanted assurances from Vickery that once the process was under way it wouldn't be aborted by City Light or anyone else. Henault had just been "burned" by the old timers in the utility over his recommendation that they respond to SEPA requirements with a "declaration of nonsignificant impacts," and he no longer viewed the utility as a trustworthy entity. He had faith in his boss, but he knew that Vickery would listen to many other City Light staff leaders.

Henault wanted a guarantee for another reason. Seattle would be having a City Council election in 1975. Two of the nine Council members were running for reelection, two who had been instrumental in having City Light investigate the need for power from WNP 4 and 5 and who were sympathetic to the environmentalists. If they were not reelected, it might be easy for a new City Council to sweep the study under the rug and tell the environmentalists to get lost. Vickery promised Henault that no one at City Light would interfere or hinder the committee's work in any way.

To see that public involvement would become a cornerstone of the study, one of the first things Henault did was write a letter for Vickery's signature. Copies of the letter went out to all who had testified before the City Council on Seattle's participation in WNP 4 and 5. The letter described Henault's ideas for the public participation program and asked for reactions. Should the study be done in-house (in City Light's offices), or should an independent outside consultant be hired? Should citizens be involved in the process or not? When the responses came back it was obvious to Henault that the people of Seattle wanted an independent study and supported citizen participation in the study.

Once public support for citizen involvement had been established, Henault and Vickery went to the City Council seeking approval of the plan and money to carry it out. The Council seemed to have no particular opinion one way or the other about citizen involvement in the study. In fact, there wasn't even much concern about whether the study was done by City Light staff or by outside consultants. Although the plan was approved, the way the Council left it, City Light could have done the study in-house in three months, with no citizen involvement. The Council did approve expenditures to hire energy forecasters and energy supply/demand consultants.

Shortly after the City Council approved the plan for a study conducted by a citizen's committee, Henault met with people from Seattle's Office of Policy Planning. They identified groups that they thought might be interested in having representatives on the committee—ratepayers, environmentalists, and the industrial and business communities. The final committee was composed of 30 people who as a group represented a cross-section of Seattle.

The committee began work in July 1975, and some tentative results started appearing in December 1975, with the final report scheduled for April 1976. Among the committee's early findings was that its projection of the need for power was only about one-fourth of what City Light's had been. The committee also hinted that it might recommend that City Light take a nonnuclear position and drop out of WNP 4 and 5. In the end, the committee went beyond its advisory role. As their work proceeded, committee members became advocates of a nonnuclear future and took their recommendations straight to the City Council, bypassing Vickery and City Light.

The City Council and the citizens of Seattle seemed pleased with the committee's work. When it came time, in April 1976, to decide on which energy path to take—hard technology, that is, participation in WNP 4 and 5, or a soft energy approach, one based on conservation—the City Council opted for the latter, by a 7–4 vote. Hence the Seattle City Council decided not to participate in the Pacific Northwest's Hydro-Thermal Power Program, Phase 2.

Vickery's own feeling was that the City of Seattle should have participated in WNP 4 and 5, with a 10 percent share, though he would have been happy with only a 5 percent share. He strongly believed that it was very important for Seattle to participate at some level in the Hydro-Thermal Power Program, and he wanted to maintain a "brotherhood" with the other utilities in the Pacific Northwest and WPPSS. If for no other reason, he believed participation would allow City Light to retain some influence over construction and operation of the HTPP projects.

In rejecting City Light's participation in WNP 4 and 5, the Seattle City Council's real issue had less to do with nuclear power than with money. The committee had determined that participation in WNP 4 and 5 would mean a substantial increase in the cost of electric power for Seattle ratepayers. The committee believed that even with a widespread energy conservation program, Seattle would need additional electric energy by the 1990s. City Light officials felt it would be cheaper in the long run to buy into WNP 4 and 5 now rather than build new generating facilities 15 to 20 years in the future.

From time to time during the committee's deliberations, WPPSS officials had come by and reviewed the committee's work. Then they would meet with City Light and laughingly say that City Light had shot itself in the foot when it established the citizen's committee to review the need for power. Yet it wasn't long before WPPSS officials were having troubles of their own in explaining to the Pacific Northwest public why the cost overruns on nuclear plants 1, 2, and 3 were escalating at the rate of $15 million a week.

5

The WPPSS Management Organization and Philosophy

In 1977, the Washington Public Power Supply System found itself committed to building five nuclear power plants at the same time, a construction program of unprecedented size in the history of the electric power industry. But that didn't seem to bother the WPPSS Board of Directors. They believed that power should be available, no matter what its cost, and that their decision to build more generating plants would result in more jobs, low electric rates, and therefore, greater social and economic stability for the people of the Pacific Northwest. Even before WNP 4 and 5 were begun there was growing evidence that WPPSS was experiencing trouble in the form of cost overruns on WNP 1, 2, and 3. These cost overruns were anticipated, the Board of Directors said, and they were certainly no reason to modify or abandon WPPSS' assigned mission to keep the region adequately supplied with electricity. By 1974, WPPSS had estimated that cost overruns on the three nuclear plants would amount to almost $1 billion. To the Board of Directors, this amount of money, unfortunate as it was, didn't change BPA's projections for regional power needs.

In 1975, the Thermal Projects Division of BPA discussed internally WPPSS' growing cost overrun problem. In December 1977, just as WPPSS was adding WNP 4 and 5 to its construction program, BPA Acting Administrator Ray Foleen began to formally question WPPSS' cost overruns and growing management problems:

> There is some concern over the 19 percent increase in owners' costs for the three Net-Billed projects during the past year. These costs approach or exceed the architect/engineers' costs for each project. There appears to be a duplication of effort between the owner and architect-engineer and several management areas, and we would like to continue working with your staff on this item.

About one month later, on January 31, 1978, BPA administrator Sterling Monroe wrote WPPSS Managing Director Neil O. Strand about the cost overruns in even stronger terms:

> As mentioned in our meeting, we are very concerned with the lack of progress [by WPPSS] over the past six months. The current costs of the Net-Billed project have a very substantial impact on our projected rate increase in 1979. Delays in the presently scheduled commercial operation dates and increased project costs will have a significant impact on BPA's resource planning and future rate studies.

Finally, in 1978, BPA Administrator Monroe, under pressure from the U.S. Senate Subcommittee on Public Works, asked the management consulting firm of Theodore Barry and Associates (TBA) to make a management study of the roles and relationship of BPA and WPPSS. TBA's report, published in January 1979, criticized WPPSS for inefficient contract arrangements with project architects and engineers, weak accounting, inadequate internal auditing procedures, an inept public relations program, and, most importantly, a management system unable to control project progress and quality. The Washington State Senate Energy and Utilities Committee (WSSEUC) conducted its own inquiry into WPPSS management operations in 1980 and reached similar conclusions. The WSSEUC also noted other WPPSS management problems such as the apparent absence of any realistic discipline in budgeting and scheduling processes, failure of the Board of Directors to adequately address policy considerations associated with WPPSS project decisions, duplicate efforts resulting from the poor management of the many contractors at any one site, and the failure to prepare cost estimates in enough detail to ensure accuracy. But above all else, the WSSEUC determined that WPPSS management had been the most significant cause of cost overruns and schedule delays on the five nuclear projects. All this criticism by and large did not detour the WPPSS Board of Directors. In July 1981 Edward Fischer, Clark County Public Utility District commissioner and president of the WPPSS Executive Committee, said there was no way he would vote to stop construction of WNP 4 and 5. Said he:

> Stopping construction [on WNP 4 and 5] would be unthinkable. The need for that stuff is so great. The cost of putting them to bed would be far greater than building them.

During WSSEUC's 1980 investigation of WPPSS' management operations, WPPSS officials maintained that they were blameless for the cost overruns. For example, WPPSS Manager of Finance Management Control, L.S. Sandlin, testified that

> 99 percent of the cost overruns were beyond WPPSS' control and represented costs due to inflation, labor strikes, low labor productivity, escalating nuclear fuel costs, other authorized costs, and constantly changing regulatory requirements.

WPPSS officials further maintained that the cost increases and schedule delays were relatively typical of the U.S. nuclear industry at large, even though WPPSS' costs reached above the overall industry average.

WPPSS Organization

The Washington Public Power Supply System was established in 1957 as a joint operating agency of the State of Washington to be responsible for planning, financing, constructing, and operating thermal electric generating plants that one public utility district or city alone couldn't afford. It was organized under the laws of the State of Washington (Revised Code of Washington, RCW, Chapter 43.52). PUDs are to provide utility services on a nonprofit, cost-of-service basis. WPPSS is composed of 19 operating PUDs and the four cities of Richland, Seattle, Ellensburg, and Tacoma.

WPPSS was initially governed by a 23-member Board of Directors, one member from each PUD and city. Its 23 members are:

Public Utility District #1 of Benton County
Public Utility District #1 of Chelan County
Public Utility District #1 of Clallam County
Public Utility District #1 of Clark County
Public Utility District #1 of Cowlitz County
Public Utility District #1 of Douglas County
City of Ellensburg
Public Utility District #1 of Ferry County
Public Utility District #1 of Franklin County
Public Utility District #2 of Grant County
Public Utility District #1 of Grays Harbor County
Public Utility District #1 of Kittitas County

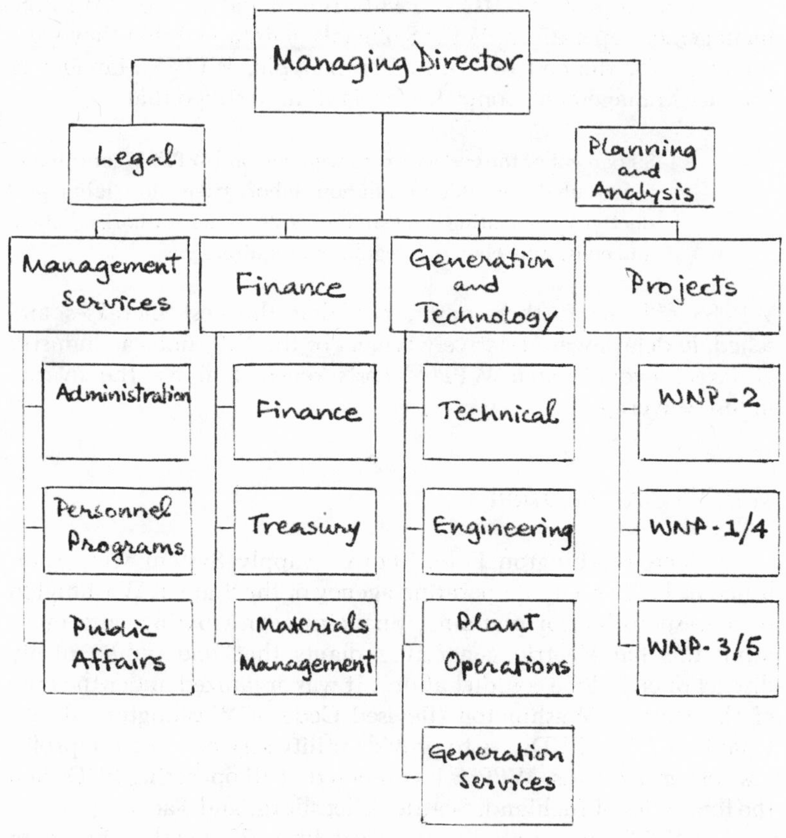

WPPSS Table of Organization, April 1, 1978.

Public Utility District #1 of Klickitat County
Public Utility District #1 of Lewis County
Public Utility District #3 of Mason County
Public Utility District #1 of Okanogan County
Public Utility District #2 of Pacific County
City of Richland
City of Seattle
Public Utility District #1 of Skamania County
Public Utility District #1 of Snohomish County
City of Tacoma
Public Utility District #1 of Wahkiakum County

The Board of Directors met quarterly. Because it was recognized that a 23-member Board was too large to be an effective reviewer of WPPSS activities, a 7-member Executive Committee elected by the full Board was formed. The Executive Committee met twice a month to act for the full Board. Day-to-day operations of WPPSS were directed by a Managing Director and an executive staff of five Assistant Directors (the directors for management services, finance, generation, technology, and projects; see Table of Organization). In addition, the Managing Director was supported by three staff functions—legal, planning and analysis, and a staff assistant.

Growth of the WPPSS staff was rapid, particularly during its later years. In 1962, WPPSS had a staff of three working in two rooms. By 1971, when work on the nuclear program began, WPPSS still had fewer than 100 employees. As late as mid-1976, the staff numbered only about 525. By late 1981, the staff had grown to more than 2,000. Typically, organizations that experience rapid growth also encounter coordination and communication problems. There is evidence that such difficulties plagued WPPSS.

The WPPSS Board of Directors

Because they had conceived and established WPPSS and also participated in the nuclear construction program, the PUDs had a vested interest in WPPSS and its projects. Because of this vested interest, the WPPSS Board of Directors for a long time was made up exclusively of representatives of its 23 members. But WPPSS did not operate in a vacuum. The PUDs in the State of Washington participated in a number of councils, committees, and associations. Each of these groups, too, had a vested interest in WPPSS' management operations. For example, the Public Power Council (PPC), which represented 114 publicly owned utilities, was interested in WPPSS' day-to-day activities because it was supposed to coordinate the public power groups in their efforts to ensure a satisfactory regional power supply. Further, the PUDs in the State of Washington were organized into the Washington State PUD Association. Finally, there was a Participants' Review Committee (PRC), which represented those who had contracted for a share of the power generated by the WPPSS plants. Thus, the group that had a strong interest in WPPSS' five nuclear projects numbered 52 cooperatives, 2 irrigation districts, 30 municipalities, 26 public utility districts, and 5 investor-

Table 3

Washington Public Power Supply System Project Status, 1982 Cost Estimate

Project	Estimated Date of Commercial Operation		Slippage	Estimated Cost, in Millions			Estimated Cost Overun, in Millions
	Original	1982		First "Rough" Estimate	First Official Estimate	1982 Estimated Cost	
WNP 1	9-1980	6-1986	69 months	$0,700	$1,204	$4,300	$3,064
WNP 2	9-1977	2-1984	77 months	$0,400	$0,500	3,200	$2,712
WNP 3	9-1981	12-1986	63 months	$1,402	$1,402	4,600	$3,130
WNP 4*	3-1982	6-1987	63 months	$1,610	$1,610	5,600	$3,900
WNP 5*	3-1983	12-1987	57 months	$1,610	$1,951	$6,200	$4,310
			329 months	$5,722	$6,657	$23,900	$17,116

Source: Senate Committee on Energy and Utilities, Energy Transition to the '80s, Olympia, Washington, 1980.

* WNP 4 and 5 were terminated on January 22, 1982.

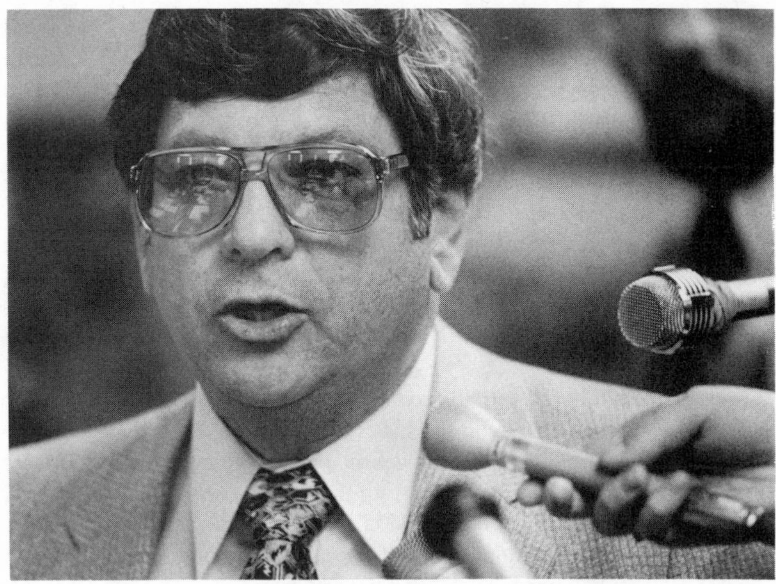

Robert Ferguson, Managing Director of WPPSS.

owned utilities. It was important that WPPSS have an effective relationship with these various public power groups. WPPSS was supposed to keep them all informed as it carried out its mission, so that the interests of public power and the ratepayers were served.

Although not cheap by hydro standards, the "first rough" \$5.7 billion price tag for the five WPPSS plants did not frighten the PUDs sponsoring the construction. (See Table 3.) Nor did it provoke general public debate like it did in Seattle. BPA said the cost of the five projects would translate into an average cost of around 3.75¢ per kilowatt-hour (a little less than residential users in the Pacific Northwest paid in 1983). But in 1977, the estimated costs for all five plants exceeded \$8 billion, and by 1980 WPPSS was saying the plants would cost a total of \$12 billion to build.

The implications of these higher costs were enormous, because the public demand for power is directly related to its cost: Like other products, as the cost of energy goes up, demand falls off. Generally, however, energy forecasters in the electric power industry in the late 1960s to mid-1970s believed that people would not be willing to cut back on electricity, even when its cost per kilowatt-hour rose, as they might cut back if the price of automobiles, movie tickets, or clothes rose.

Therefore, when WPPSS and the PUDs decided to go ahead with the construction of five nuclear power plants, their decision was based on BPA's estimate of a continuing 7 percent growth in electric consumption. Apparently BPA energy planners didn't believe for a minute that electricity consumption would ever fall off, and neither did the PUDs. In the late 1960s BPA kept stating that the Pacific Northwest was going to be short of electric energy by the mid-1980s, even with the completion of all five WPPSS nuclear plants. Robert Ferguson, Managing Director of WPPSS, speaking for the entire WPPSS organization, outlined the seriousness of its mission:

> What happens to the economics and really the social stability in a region where somebody has to make a decision on who gets the power and who doesn't get the power? To my knowledge, there is no institution that is set up that could say hospitals get power, aluminum companies don't get power. Who makes those decisions? I don't know if people understand that that's what would happen if we didn't have power.

This explains why WPPSS management and its Board of Directors believed that the only way to meet future energy demand in the Pacific Northwest was to build more plants, that without adherence to the WNP construction program (despite escalating costs) the region would be unable to avoid energy deficits in the late 1980s or sooner. BPA, the PUDs, and WPPSS management seemed to believe they were embarked on a holy construction crusade to save the Pacific Northwest from the horror of electric energy shortages. Nothing, not even the doubling or redoubling of the price tag, would detour them from fulfilling their mission. So seriously did they regard their cause that when the Seattle City Council voted to emphasize conservation, WPPSS officials and supporters isolated Seattle City Light from its public power brethren for years. (Even today, the region's public utility officials still grumble over City Light's decision to do without WNP 4 and 5.)

In the mid-1960s, there was no feeling in the Pacific Northwest that utilities should base energy planning on public input. BPA drew up the HTPP and carried out the planning and construction program with WPPSS in relative isolation from any public energy/environmental debates so common in the 1970s. Public questioning was discounted—or considered almost subversive. Said BPA Administrator Donald Hodel in remarks before the Portland (Oregon)

City Club in 1975, when all five WPPSS nuclear electric projects were under way and environmental groups were questioning projections of power demand in the Pacific Northwest:

> The environmental movement has fallen into the hands of a small, arrogant faction which is dedicated to bringing our society to a halt. They are the anti-producers, the anti-achievers. The doctrine they preach is that of scarcity and self-denial. The prophets of shortage are keenly aware that our economic well-being is closely linked to the production of energy. By halting the needed expansion of our power supply system, they can bring this region to its knees. I can tell you they are well on the road to accomplishing that goal.

After leaving BPA, Hodel went on to become an aide to U.S. Interior Secretary James Watt and then to become Secretary of Energy.

Experience of WPPSS Directors

Since WPPSS' formation in 1957, management and control of the organization had been in the hands of its Board of Directors. The Directors had been drawn from the member PUDs and municipalities. These individuals, largely part-time volunteers, represented the interests of their fellow ratepayers and were responsible for setting the policy that guided WPPSS' activities. In general, these Board members were leaders in their PUDs and their communities. They came from a variety of occupations, but their backgrounds were remarkably similar — small wholesale and retail businessmen, farmers and ranchers, PUD commissioners or superintendents — and they had unquestioned faith in contemporary forecasts of electricity demand and in the conventional wisdom that nuclear power was the wave of the future. Former WPPSS Managing Director Neil O. Strand summed up WPPSS' philosophy on nuclear energy: "It was getting to the point where we needed power at any price." Had they continued to work in a static environment of mature, tested technology like hydroelectric dams, the WPPSS Board might have been okay. The nuclear electric power industry was a different environment, however — it was dynamic, involving sophisticated contractors, militant labor unions, and constantly changing safety regulations.

The skills needed to make informed decisions in the expensive, fast-moving and rapidly changing world of facilities as huge and complex as nuclear power plants are not found on a Board of Directors composed solely of small-time businessmen and local public utility district members. For example, no WPPSS Directors had backgrounds in such important matters as investment banking, heavy industry, nuclear engineering, economics, and corporate management. Nick Cain, President of the Board of Directors, was an orchardist from the small town of Malott in Okanogn County. Ed Fischer, an original member of the Board and the first and only chairman of the powerful Executive Committee, was a retired salesman of small electrical appliances. Some people argue that important skills could have been obtained by hiring consultants, but the truth is that a consultant's report is no substitute for a vote.

The Decision-Making Process: Closed

Most substantive discussions of WPPSS policy took place in the Participants' Review Committee, which was composed of 7 individuals representing the 115 cooperatives, municipalities, public utility districts, and investor-owned utilities in Oregon, Washington, Idaho, Montana, California, Nevada, and Wyoming that were participating in WPPSS projects. After a decision by the Participants' Review Committee a matter would be brought before the Executive Committee and then to the Board of Directors, for endorsement—and often it was given with a simple voice vote and little or no explanation or discussion. Likewise, most major decisions by the Executive Committee and the Board of Directors required approval by the Participants' Review Committee. Even WPPSS' Board of Directors, says former Board member C. Stanford Olsen, was nothing but a "rubber stamp" for WPPSS management and the Participants' Review Committee.

From WPPSS' beginning, its Board of Directors, Executive Committee, and Participants' Committee operated pretty much behind closed doors. Critics say that WPPSS' decision to proceed with the first three nuclear projects under BPA's net-billing financing plan was known to relatively few people in the Pacific Northwest outside formal power planning circles. Likewise, WPPSS' ill-fated decision to finance WNP 4 and 5 through a capitalization of interest plan was equally unknown to most everyone outside the Participants' Review Committee. All the power plant planning was done during meetings of BPA officials and utility executives, to which the PUD membership and the general public were not invited.

Like most states, Washington State has laws pertaining to public meetings and records. The Open Public Meetings Act reads:

> The people of this state do not yield their sovereignty to the agencies which serve them. The people, in delegating authority, do not give their public servants the right to decide what is good for the people to know and what is not good for them to know. The people insist on remaining informed so that they may retain control over the instruments which they have created.

In August 1981, after hearing wave after wave of criticism, Participants' Review Committee Chairman Edwin Morris, a delegate from the Clark County PUD, sought to overturn the Committee's secrecy policy. Morris' motion was defeated, and he resigned his post after Committee members accused him of "grandstanding." A month earlier a Seattle resident had gone to court to force the Participants' Review Committee to meet publicly, but King County Superior Court judge had ruled that WPPSS' Participants' Review Committee was not a public agency and could freely meet in private. Following the mid-August confrontation between Morris and his committee colleagues, Washington State Representative John Erak (R–Aberdeen) vowed to introduce legislation that would force the Participants' Committee to meet in the open. But that tactic had been tried before, to no avail: The Washington Legislature had passed a law in 1979 requiring the Review Committee to meet publicly, but the measure had been vetoed by Governor Dixy Lee Ray.

Glen C. Walkley, long-time commissioner of the Franklin County PUD, said in an interview with the *Oregonian* newspaper that WPPSS in general resented outside interference of any kind by anybody. Walkley should know: as a member of the Participants' Committee and for 23 years a member of WPPSS' Executive Committee, no one had more to do with the formation, functioning, or philosophy of WPPSS than he did.

Walkley owned a 5,000-acre wheat farm fronting the Columbia River. His farming success was due, he said, to electricity. Before electricity was available, the area was essentially a desert. To Walkley, electricity was more than the stuff of progress. It was the stuff of civilization itself.

Walkley believed that he was elected to WPPSS by the people of the Franklin County PUD to use his own best judgment in decisions on power supply and energy policies. Although he did acknowledge that many important decisions were made for Franklin County PUD ratepayers without consultation, he was not an advocate of citizen participation. When other members of the Franklin County PUD board proposed a creation of a citizens advisory board, he fought it and won. "We are all farmers. We all belong to the Grange, the Elks, the Lions, the Chamber. If anyone wanted to talk to us, they got us there," he said about citizen participation. He resented any outside interference in WPPSS affairs. "A lot of the WPPSS critics," he said, "are people who aren't even ratepayers. A lot of them are people who maybe don't even buy electricity."

Walkley also resented the intrustion of the Washington Legislature into WPPSS' business. Said he: "I think a great deal of the trouble has been the legislative committee [a committee organized by Senator King Lysen seeking to open up the Participants' Review Committee and the WPPSS decision-making process in general]. Would [the legislative committee] know more about this organization than I would, the man who went to all the meetings?" "The people who buy and build the projects should have control over them," he said. "If they would lay off and let us alone, [the nuclear plants] would be cheaper." At the end of Walkley's interview, he offered his philosophy about public affairs, which helps explain why the WPPSS Participants' Review Committee meetings were, as a matter of policy, closed to the public unless expressly opened by a vote of the participants:

> When you are talking about a lot of these scientific things, you don't want to jaw about things with [the public] all day. We're not going to argue back and forth with people. It takes all day. I don't care what it is, when you are talking about something new, you are going to have a certain proportion of again'ers.

6

The BPA Role and Relationship with WPPSS

Initially, the Bonneville Power Administration was the only organization that made energy forecasts for the entire region. Consequently, it played an important role in helping the public utility districts plan additional generating facilities. This assistance varied from providing minor help on technical issues to actually preparing demand forecasts for the smaller PUDs. PUD forecasts were sent to BPA headquarters for review and use in the regional energy forecasts. Final approval of the forecasts rested with the PUDs. However, most of the PUDs were small utilities with limited staffs and limited technical expertise in demand forecasting. As a result, they relied on BPA's expertise and accepted the forecasts prepared by BPA.

BPA also made demand estimates for Federal and State agencies and for the industries it served directly. As a result, the PUDs, Federal and State agencies, and direct service industries endorsed and agreed with BPA's belief that additional generating units were needed to meet regional power needs.

BPA not only did most of the forecasting that called for additional generating units, but also publicly defended the forecasts against critics. In response to a critique from the Washington State Thermal Power Plant Site Evaluation Council, BPA stated that the forecast was the "best available tool for projecting the future demand for electricity in the Pacific Northwest." BPA also expressed concern that the regional forecast might, in fact, be too low. The need for generating capacity was beyond the capabilities of any one PUD. Therefore, WPPSS would construct five nuclear power plants. Through the net-billing agreements, BPA would provide financial backing for three of the five plants and to play its traditional role of marketer for the output of all the plants.

Although BPA made the regional forecasts and, thereby, the power deficit projections, from data collected from the PUDs, there is no indication that BPA made an independent evaluation of the methodology or assumptions used by the PUDs. A possible reason for this was that the demand growth predicted for the Pacific Northwest, about 6 to 7 percent, was similar to projections for the rest of the Nation.

BPA's Oversight Role

By virtue of its Net-Billing Agreements with WPPSS BPA has a legitimate oversight role in WNP 1, 2, and 3. Under the Net-Billing Agreement terms, BPA is obligated to pay Participants for the total annual project costs, whether or not the projects are completed or operable. In addition, BPA is obligated by its Federal charter to recover, through charges to customers, all its construction expenditures and operating and maintenance costs, including debt service on bonds issued to finance construction of WNP 1, 2, and 3.

Working Relationships

The working relationship between BPA and WPPSS during construction of the five nuclear electric generating plants was defined in the Project Agreements between BPA and WPPSS. Although the Agreements did not specify an oversight role for BPA, the intent did seem to be to give BPA a mechanism for ensuring that the interests of BPA customers were protected. However, the role of BPA, specifically its Thermal Projects Personnel, remained confused.

Over the years, several attempts were made by BPA and WPPSS to achieve an acceptable working understanding of BPA's role in the construction of WNP 1, 2, and 3. However, none resulted in a satisfactory solution. The WPPSS Board of Directors wanted BPA to have a significant oversight role: Because of their technical strength, BPA staff could provide an effective check and balance on WPPSS actions. The WPPSS staff, in contrast, believed that any kind of oversight was only an "out of context" exercise, that meaningful insight into project management could not be developed without detailed day-to-day involvement.

Most of BPA's oversight function on WNP 1, 2, and 3 was performed by a project engineer at each of the three sites. These project engineers monitored activities at the sites and periodically reported on the project's progress to BPA's Manager of Thermal Projects. This monitoring activity consisted of several elements, including:

Review of documents supplied by WPPSS;
Informal analysis of project progress trends;
Informal meetings with project management personnel to discuss project progress; and
Review of items requiring BPA approval, such as the budget or contracts and change orders over $500,000.

Little Interaction

Throughout much of the 1970s there was very little regular interaction between the senior management of WPPSS and BPA. Privately, WPPSS' management felt they were spending too much time answering questions from BPA personnel. This lack of satisfactory interaction led to a deterioration in confidence between the two agencies. Under these circumstances, it was somewhat difficult to maintain the partnership necessary for on-time and within-cost completion of these complex projects. Instead of getting the information it felt was necessary to help ensure "on-time" and "least-cost" construction on WNP 1, 2, and 3, BPA received copies of correspondence between WPPSS and its construction engineers and contractors and between construction engineers and the contractors, and copies of some management reports. Certainly this was a lot of information—too much, really, and it was not summarized so as to be useful for quick analysis of project status. In addition, the information was not complete. BPA didn't receive all internal management reports, so it was not fully aware of the total project control efforts and project status. For example, BPA did not receive internal audit reports, and, consequently, it was not completely aware of the effectiveness of the internal auditing group. The result was that BPA was not satisfactorily informed about project progress and corrective action required to achieve "on-time" and "least-cost" project completion.

WPPSS management did not always respond to BPA's requests for information on a timely basis. For example, during the 1979 budget review of WNP 2, BPA requested copies of cost-effectiveness studies that supported the initial and continued schedule acceleration decision. Satisfactory information was not made available to BPA prior to the time for budget approval. Thus, it was difficult for BPA to make an informed judgment about the schedule acceleration costs and benefits.

If, as a result of the information provided to BPA by WPPSS, BPA personnel had any comments or recommendations, they had to communicate them to the WPPSS staff. BPA found that their comments and/or recommendations were rarely passed on to the Board of Directors or the Executive Committee. Consequently, BPA found that it wasn't very effective as a "check and balance" mechanism, and what it could accomplish was diluted.

Lack of BPA Leverage in Its Oversight Role

One would think that BPA, since it in effect guaranteed WPPSS bonds on WNP 1, 2, and 3, would make every effort to ensure that the money was well spent, but this was not the case. BPA did little to strengthen its oversight position. There are several reasons for lack of action:

Seeking more influence over WPPSS would have been a lengthy, time-consuming process;

The act of seeking more influence would have damaged the partnership spirit between the two organizations;

A struggle over seeking more influence would have affected the public image of both organizations;

Seeking more influence would have affected the project schedule directly or indirectly, thus also affecting costs.

BPA did not have a significant oversight relationship with the WPPSS Executive Board or the Board of Directors. The Board members felt uncomfortable both in a decision-making role and because of their limited technical and management experience in the nuclear power plant field. The information each Board received from management was of little help in establishing a clear understanding of the overall WPPSS situation. They had difficulty knowing what questions should be asked of their WPPSS management. Consequently, BPA found it difficult to perform their oversight role with Board members with any degree of satisfaction.

7

The WPPSS "Capitalization of Interest" Financing Scheme

The WPPSS management and its Board of Directors have been blamed for most of the problems that have plagued the ill-fated nuclear program. As easy and convenient a target as WPPSS' management and Board of Directors may be, they are not the real culprits. In fact, it is questionable whether anyone could have brought WPPSS' five-plant construction program to completion on time and within budget, especially given the method of financing WNP 4 and 5.

WPPSS' Financial Planning

WPPSS, as a municipal entity, was authorized to finance its nuclear electric generating program through tax-exempt bonds. Its chief funding source was long-term bonds—that is, bonds having a life longer than six years. For such bond issues, the Washington State Auditor was responsible for authenticating the bonds; the WPPSS Board of Directors and BPA were responsible for approving all WPPSS financing. WNP 4 and 5 were to have been handled as WNP 1, 2, and 3 had been, with financing guaranteed by BPA through Net-Billing Agreements. Then came the IRS regulation that if tax-exempt bonds were used to finance energy projects, not more than 25 percent of its output could be purchased by a Federal agency. This meant that overall financing requirements for WNP 4 and 5 had to be guaranteed by the 88 Participants.

So as not to alarm the participating utility ratepayers, who might be unwilling to see their electric rates rise during the construction years, WPPSS decided on a finance program that depended on the issuance of revenue bonds. These bonds would pay not only the capital costs of construction, but also the interest accrued on construction funds during the construction period. This "capitalization of interest" approach was a giant gamble for WPPSS. It would benefit ratepayers in the participating utilities since their rates would not go up until power was produced. On the other hand, it would greatly increase the amount of debt that ultimately must be issued and, consequently, the debt service charges when the plant finally came into production. If interest rates rose, the ultimate cost of the project would increase dramatically because high interest charges during construction would add to the principal amount on which high interest must be paid.

Capitalization of interest is terribly risky business. Few companies in the electric power industry use this means of financing nuclear electric projects. Investor-owned utilities, for example, finance their construction projects from current earnings; all costs of work in progress are passed on to ratepayers in the form of higher electric rates, and thus electric consumers pay for construction costs as the bills become due. According to TVA Chairman of the Board S. David Freeman, "The difference in costs between TVA's nuclear projects and WPPSS' [$3 billion each versus $6 billion each for WPPSS]" said Freeman in 1981 "is due to capitalizing the interest." TVA had rejected capitalization of interest because the scheme was considered financially irresponsible.

WPPSS officials knew there were some risks but decided to finance WNP 4 and 5 on a giant "charge card" anyway. It was a gamble WPPSS believed it could win because of its previous experience in building electric generating facilities on time and within budget. Officials never really felt that building a nuclear power plant was much different from building a hydroelectric project like Packwood Lake. Nor were they worried by all the industry talk about the complexities and the delays which have come with nearly all nuclear power plant construction projects. The thing that did worry them was that a lack of money for construction had caused a great many woes for the utilities sponsoring nuclear electric power plants. This would not happen to WPPSS, they reasoned, because their "capitalization of interest" plan would ensure that construction schedules were not upset by a shortfall in funds. This practice could be safely carried out, WPPSS officials reasoned, because of their easy access to "unlimited," low-cost, tax-exempt financing.

As they got into their five-plant nuclear construction program, WPPSS officials found that they, too, were experiencing numerous setbacks in their plant completion-operation timetables. As construction schedules slipped, they saw their overall costs increase — and would later find that their "capitalization of interest" scheme was, in effect, "bleeding them white" as the bond markets for their paper became saturated and unresponsive. Unable to borrow more money through bond sales to both pay the huge interest debt and also keep construction going, WPPSS officials felt certain that if the 88 Participants would help them pay at least the interest due on previously borrowed money, then the bond market would again readily purchase their tax-exempt bonds. But the Participants said "no," thus forcing WPPSS to abandon the partially completed WNP 4 and 5.

It appears that the real culprit in the WPPSS tragedy, if there is one, is someone's initial belief that WPPSS could bring five nuclear electric generating plants into operation on schedule, when other nuclear plant owner/builders across the Nation couldn't do it with one or two plants. WPPSS lost its gamble with the "capitalization of interest" scheme for WNP 4 and 5 and as a result left several hundred thousand investors holding $2.25 billion of worthless bonds.

Why did WPPSS take the risk of capitalizing the interest? In 1973, interest rates and construction costs were much lower than they were when WNP 4 and 5 were terminated in January 1982. (In June 1973, for example, WPPSS' first bond offering of $150 million was sold at 5.65 percent interest. WPPSS' last bond offering, in September 1983, for $750 million on WNP 1, 2, and 3, sold at an interest rate ranging from 14.25 percent to 15 percent.) WPPSS' plan called for the 88 Participants to make no payments until the plants went on line in the mid-1980s and were generating revenue through electricity sales. Until that time, WPPSS would set aside enough money from every bond issue to pay interest on the bonds for two years. (For the final bond issue for WNP 4 and 5 in 1981 for $200 million, for instance, WPPSS set aside $101.2 million for interest payments and other reserves; only $90.4 million was earmarked for construction.) As long as the nuclear projects arrived at successive milestones on schedule and were finished on time, there would be no financial dangers for WPPSS or its 88 Participants. Even cost overruns could be tolerated.

However, delays caused by labor unrest, low labor productivity, lateness in the delivery of equipment and supplies, and safety modifications required by the Nuclear Regulatory Commission began pushing operating dates further into the future. This state of affairs couldn't be tolerated for long, because it meant revenues needed to cover the interest on money already spent would not be

Table 4

Annual WPPSS Debt Sales, 1973–1981, in Millions of Dollars

Year	*WNP 1, 2, 3*		*WNP 4, 5*		*Total*	*Cum. Total*
1973	150		0		150	150
1974	230		0		230	380
1975	450		100		550	930
1976	780		0		780	1,710
1977	230		365		595	2,305
1978	740		470		1,210	3,515
1979	455		575		1,030	4,545
1980	410		690		1,100	5,645
1981	425		200		625	6,270
1982	850		0*		850	7,120
	4,720	+	2,400	=	7,120	

*WNP 4 and 5 were terminated January 1982.
Sources: Bond Buyer's Municipal Financial Service, New York; The Bond Buyer, various editions.

---

forthcoming. WPPSS would have to sell more bonds just to cover interest due on previous bond sales. In 1981 WPPSS officials need to raise twice as much money in the bond market as they did in 1980.

Capitalizing interest costs on its nuclear electric program had by November 1981 left WPPSS with a $6.2 billion debt (see Table 4). All five nuclear projects were now 500 percent over budget and up to six years behind schedule. WPPSS had a voracious financial appetite requiring regular "feedings" merely to pay interest on its debt, not to mention monies for its construction costs. In September 1981 the Washington State Governor's Panel Report noted:

> Of the proceeds of debt issuance for WPPSS WNP 4 and 5 in the next two years, less than one-half would be available for construction. The majority would be required to pay interest on bonds already issued, along with financing costs and amounts held in reserve. During fiscal year 1984, only one-third would be available for construction. During fiscal year 1985, one-fifth would be available for construction.

According to WPPSS' 1982 Fiscal Year Budget, some $9 billion in debt remained to be issued for the completion of WNP 4 and 5, and an additional $4 billion remained to be sold on WNP 1, 2, and 3, for a total of $13 billion by the expected final completion dates of 1987–1988. The total cost for completing all five nuclear projects was now projected at $23.9 billion. In addition, WPPSS budget documents reveal how expensive it was becoming to bring their bond issues to market. WPPSS paid more than $23.7 million in fees and commissions to the brokerage houses that handled its record $850 million bond sale in February 1982.

The sheer amount of money needed was getting out of hand, and Wall Street was becoming nervous over the slipping schedules. The brokers began telling WPPSS that WNP 4 and 5 bonds were becoming close to "unmarketable" because of downgrading by major bond rating agencies and warnings to investors by investment houses about the soundness of the bonds.

Saturation Affects All of WPPSS' Debt

By 1982, WPPSS' financing requirements were so great that the ability of the market to absorb it all was being strained. Noting that WPPSS would have to borrow up to $13 billion additional to finish all five projects in 1988, personnel at Merrill Lynch stated: "This level of financing would be unprecedented and would exacerbate the already evident WPPSS saturation of institutional investment portfolios." In terms of survival of WNP 4 and 5, WPPSS had essentially put itself at the mercy of the bond market. Not only was there a limit to what the financial community would tolerate, but there was also a limit to what the public would tolerate.

WPPSS' $6.8 billion bonded indebtedness in 1981 represented approximately 2 percent of the total in municipal bonds outstanding ($348.6 billion), according to Salomon Brothers of New York. On the surface, this doesn't seem like saturation, but an examination of the investor profile puts saturation in a different perspective.

Although the tax advantages of municipal bonds are attractive to many investors, the tax-exempt privilege is not important to some investors. Pension funds and life insurance companies, for example, have little need to invest in them because they already have tax advantages.

The largest owners of municipal bonds, according to Salomon Brothers, are commercial banks, insurance companies, households (that is, individuals), and municipal bond funds. In the early 1980s, the economic climate was not favorable for large municipal bond purchases by banks and casualty companies. Bank earnings had been skimpy, so banks did not need tax-exempt investments. Moreover, banks were leaning toward industrial development bonds for their tax-exempt investments. Casualty companies were faced with underwriting losses. The early 1980s was a bad time to expect help from these two sources. That left households, mainly, and municipal bond funds. But there is a limit to how much they can invest, especially the households. Another factor affecting the sale of WPPSS bonds was the restrictions institutional investors have on how much of their portfolios can be invested in municipal bonds, and even on how much can be invested in bonds from a particular geographical region.

The saturation point falls if for any reason the credit quality of a bond issue deteriorates. If there are significant management problems, such as cost overruns and construction delays, confidence in the borrower diminishes and the perceived risk associated with its bonds increases. This results in lowered bond ratings and the lowering of portfolio saturation levels among large investors.

In WPPSS' $200 million bond sale for WNP 4 and 5 in March 1981, $101.2 million had to be put aside for interest payments and other reserves during the 30-year life of the bonds; only $90.4 million was earmarked for construction. Interest payments typically being deferred for two years, WPPSS would have to start paying interest in March 1983. Making those interest payments would depend on WPPSS' ability to sell additional bonds. To make sure WPPSS could continue selling bonds even if saturation levels were reached, WPPSS officials and their financial planners added a "sweetner," an extra inducement to make it easier to market the bonds. In their $200 million bond issue of March 1981, a total of $30 million was to be payable, at the holder's option, on July 1, 1991, or anytime thereafter. Thus it was possible that WPPSS would have to pay off as much as $30 million in 10 years. That would have been good for bondholders, but not necessarily good for the issuer, which would have had to come up with the money.

In spite of the inducements WPPSS was offering, investor confidence was steadily softening. Investors were uncertain about WPPSS' ability to complete WNP 4 and 5 anytime soon and to sell additional bonds. In fact, it appeared that many investors, rather than buying, were selling WNP 4 and 5 issues – and advising others to do likewise. Furthermore, bond dealers were not willing to attempt to sell a new issue of WNP 4 and 5 bonds because of possible legal liability for encouraging investors to purchase excessively risky securities. Despite their reasonably high rating (Baa-1 by Moody's), WNP 4 and 5 bonds were not considered even "medium grade securities," that is, securities that bond dealers and managers of large portfolios could recommend and purchase without opening themselves up to legal liability for violating fiduciary responsibilities. This lack of "quality" was later reflected in Moody's January 1982 suspension of its Baa-1 rating and Standard and Poor's inclusion of WNP 4 and 5 issues on its "Credit Watch" list.

"Take or Pay" Contracts Untested

The downgrading of WNP 4 and 5 bonds clearly weakened investor confidence in WPPSS' ability to complete the projects and pay off its huge debt. Moreover, the "take-or-pay" contracts (also known as "hell or high water" contracts) being used with WNP 4 and 5 had never been tested in court. The "take or pay" contract was devised by natural gas pipeline companies and has become a standard feature of large-project finance. It obliges the customer to "take" the project, in this case electric power, as soon as it is produced, and to pay for it after an agreed-upon date even if no product is delivered. Strange as it may seem, take-or-pay contracts had never been seriously tested in the courts because no big project had ever failed as WNP 4 and 5 were about to do. Some bond dealers felt that the WNP 4 and 5 bonds might not be paid off if the projects failed to produce any electricity, despite the exacting take-or-pay wording in the Participants' Agreements. Their growing concern also threatened the marketability of all WNP 4 and 5 issues.

If WPPSS were to finance further work on WNP 4 and 5 through the municipal bond market, the perceived risk had to be eliminated, or at least reduced to levels that would permit bond traders to accept WNP 4 and 5 as "medium grade" and then sell to a broader spectrum of the market. To reduce the risk, a number of things would have to be done, some brokers said:

1. Studies would have to be undertaken to determine that, despite the expected relatively high cost per kilowatt-hour, there would be electrical demand in the Pacific Northwest to make WNP 4 and 5's output salable.

2. The 88 Participants and their ratepayers would have to share more visibly in the risks of WNP 4 and 5 by beginning to pay at least some of the interest. (Paying interest during construction would have increased BPA's and/or the 88 Participant's role in WPPSS management activities, possibly making it more cost effective and decreasing the amount WPPSS needed to borrow.)
3. The financial base of the 88 Participants would have to be expanded to ensure payment of the bonds in the event no electricity was produced by the two projects. (This might be accomplished by obtaining a net-billing arrangement with BPA similar to that for WNP 1, 2, and 3, establishing a partnership among the 88 Participants, the investor-owned utilities, and the direct-service industrial customers to share costs more equitably, and/or locating other partners, possibly within Washington State government agencies, investor-owned utilities of California, even the government of California.)
4. The managers of WPPSS would have to show evidence that they were capable of controlling costs and schedules, completing the construction of the plants, and producing power at cost-effective rates.
5. The uncertainties surrounding Washington State voter Initiative 394 (passed in November 1981; see Chapter 10) would have to be resolved.

Not all financial advisors agreed that all of these conditions would have to be met before additional WNP 4 and 5 bonds could be sold. In fact, the financial community was not in agreement about exactly what conditions must hold for a "medium grade security." It was clear, however, that if they were to reenter the market, WPPSS management and the 88 Participants would have to make substantial progress in several areas.

For a time it appeared that help might be on the way. In August 1981, WPPSS' Participants' Review Committee agreed to assume 50 percent of the interest during construction, with payment to begin in 1983. However, a number of the 88 Participants balked at paying any interest. Their reasoning was that when negotiations on WNP 4 and 5 were going on in the mid-1970s, they had assumed that these projects would be net-billed like WNP 1, 2, and 3. When the IRS had forced BPA to give up net-billing arrangements for WNP 4 and 5, the 88 Participants' claim, BPA had told them: "Go ahead and sign anyway. We'll take care of you after the case [with IRS] is settled."

The time for a life or death decision on WNP 4 and 5 was fast approaching. The money would be gone in March 1982. After lengthy debates, the 88 Participants finally agreed in January 1982 that they would make payments on the interest—if the previous financing and the future construction money needs were restructured. This was impossible, and WPPSS had no choice but to terminate WNP 4 and 5 immediately.

8

The WPPSS Is Accused of Mismanagement

In the early 1970s, when Washington Public Power Supply System decided to build five nuclear plants at once in Washington State, the total "first rough" estimated cost was $5.7 billion. By 1982, the figure was $23.9 billion, a staggering 583 percent increase. What was the reason for WPPSS' enormous cost overrun? A nine-month inquiry into WPPSS' management by the Washington State Senate Energy and Utilities Commission in 1980 concluded that "mismanagement was the most significant cause of the cost overruns and schedule delays." Mismanagement has also been cited by WPPSS' critics as the reason for such problems as shoddy workmanship, poor quality control, work stoppages, low worker morale, poor safety practices, and faulty construction management. These problems, say the critics, contributed billions of dollars in cost overruns.

Can WPPSS management, as problem plagued as it was, be blamed for all of WPPSS' $19.8 billion cost overrun? No, but it may be years before everything about the cost overruns is known and understood. From the evidence currently available it appears that WPPSS management initially concealed the actual costs of their five projects by as much as $5 billion. The reason WPPSS officials did not want to scare off Wall Street money lenders whose cooperation they desperately needed for their "capitalization of interest" financing scheme. It appears now that WPPSS' debt service—that is, interest—may have contributed up to 50 percent of the cost overruns. WPPSS officials believe that about 10 percent of the five-project cost overruns may have been due to management's liberal cost-control policies with their equipment and material suppliers, construction contractors, labor, legal counsel, bond underwriting fees, project site-area communities, and services. The remainder of the cost overruns, says WPPSS, may have been due to changing safety regulations, State laws that required management to accept the lowest bid for services and materials regardless of whether or not the bidder was qualified, cost-plus contracting, and, finally, inflation.

The Megabuck World of Nuclear Plant Construction

In citing "mismanagement" as the chief reason for WPPSS' troubles, many critics have singled out the WPPSS Board of Directors. Critics say the 23-member, male-only Board, composed of representatives of 19 participating public utility districts and four municipalities, was utterly lost in the megabucks world of nuclear power plant construction. Because of the directors' inexperience, critics note, the Board was always at the mercy of highly sophisticated and greedy labor unions, suppliers, and general contractors. The prolonged struggle between WPPSS' contractors and their labor unions caused WPPSS to sustain more than $1 billion in losses due to strikes and labor disputes. One four-month strike at WNP 1, 2, and 3 alone, in the summer of 1980, added an estimated $707 million to the costs of the three net-billed projects, according to WPPSS' own estimates. How much can the Board of Directors be blamed for labor problems between the contractors and the unions? What about the management, the people who ran operations on a day-by-day basis? Is there any hard evidence that WPPSS management failed in its job building five nuclear electric projects at one time?

In January 1979, Theodore Barry & Associates (TBA) made public the results of its Bonneville Power Administration-financed management study of WPPSS' three net-billed nuclear projects. TBA had studied the effectiveness of WPPSS' organization, its project management, engineering and contracts management, financial and accounting controls, quality assurance, records management, and operations planning. TBA found that WPPSS had perhaps one of the greatest pools of technical talent in the nuclear power plant industry but it did not really address the reasons for WPPSS' cost overruns. It did note that the current estimates for installed costs of WPPSS' three net-billed projects appeared to be higher than typical for nuclear power projects of the same vintage elsewhere, though they were within the upper limit of the range for other projects (see diagram on page 67).

TBA pointed out that its comparison of WPPSS' installed cost per kilowatt with the nuclear industry as a whole was only somewhat useful for defining the relative performance among many

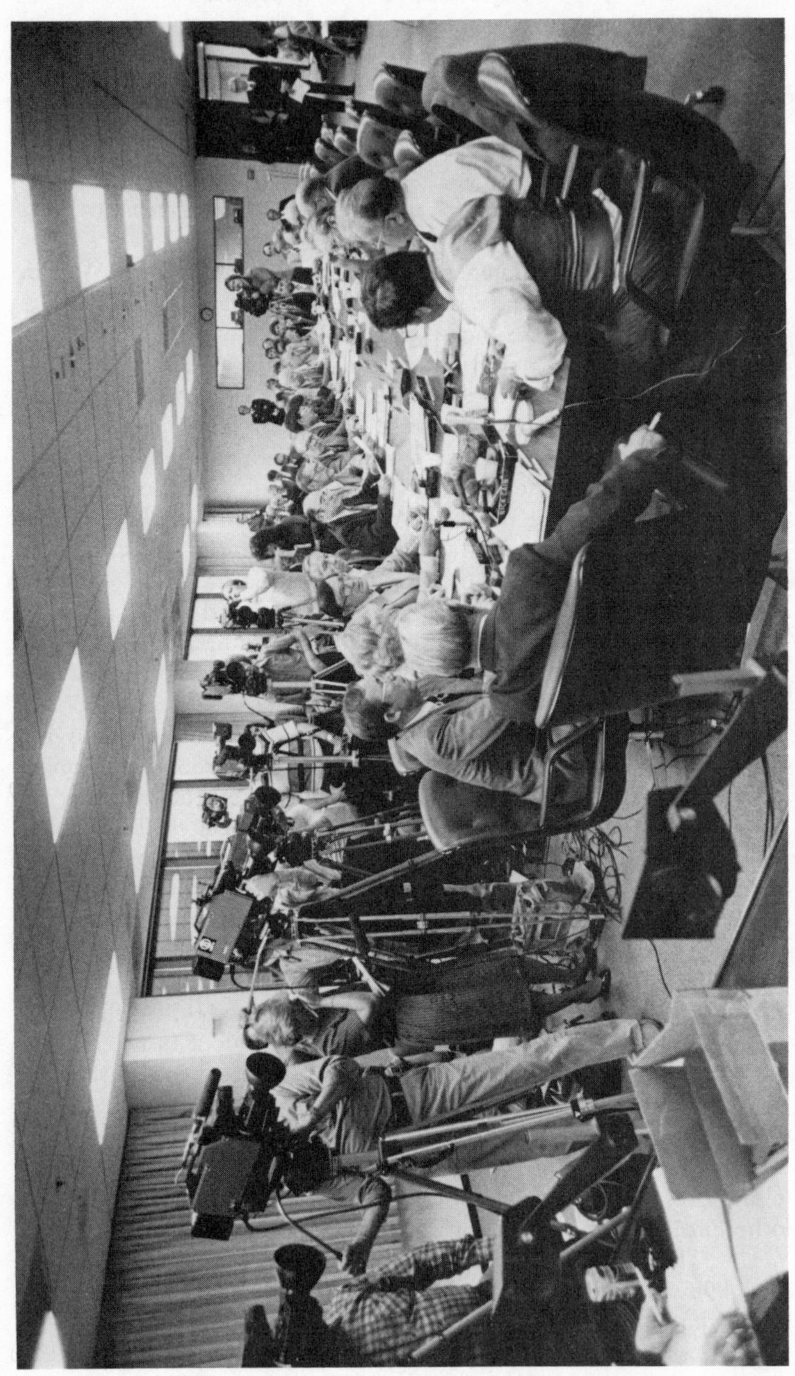

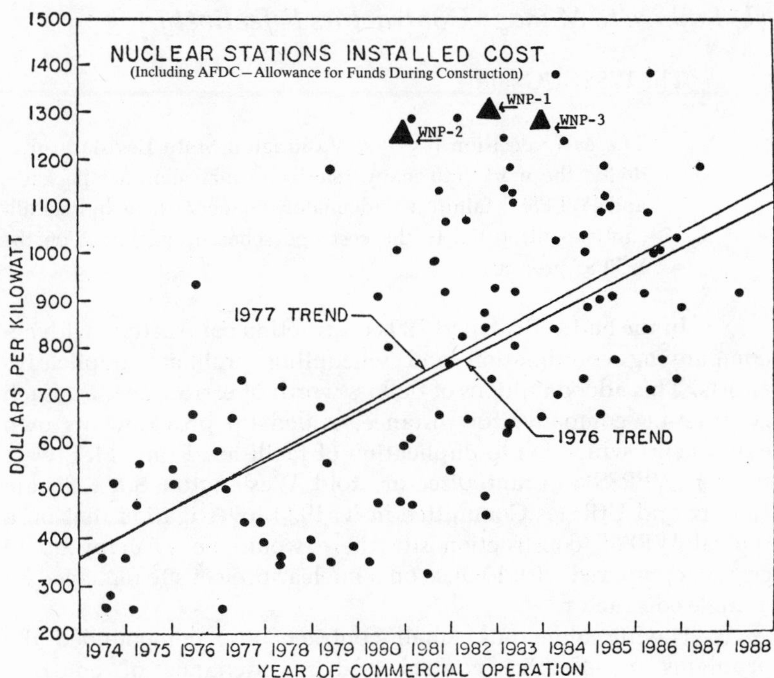

projects, because there are many differences among electric companies, such as the type of financing, the interest rates paid, the effects of inflation, safety regulations, labor productivity, and contracting practices.

Too Many Bosses and No Bosses

Regardless of what TBA found, critics have maintained that WPPSS caused its own massive cost overruns by its handling of contractors and labor unions, and its liberal attitude on spending money borrowed at extremely high interest rates. The Washington State Senate Energy and Utilities Committee study of WPPSS' schedule delays and cost overruns concluded that WPPSS managers were guilty of a number of major deficiencies, all of which contributed mightily to cost overruns. The major management deficiencies cited were:

Opposite: Typical WPPSS Board of Directors meeting, circa 1981.

(1) Failure to Manage Contractors Effectively

The WSSEUC said:

> The early decision [by the Washington State Legislature] to divide the work into many relatively small contract packages and [WPPSS'] failure to adequately manage them has significantly contributed to the cost and schedule problems on the WPPSS projects.

In the end WPPSS had 720 construction contractors, with accompanying coordination and scheduling problems duplicative efforts. This added millions of dollars worth of extra expenses. Each contracting company, for instance, ended up providing its own equipment, which led to duplication of facilities. Frank McElwee, former WPPSS assistant director, told Washington State Senate Energy and Utilities Committee in its 1980 investigation that on a typical WPPSS construction site there would be as many as 40 cranes, compared with 10 or so on a nuclear project site managed by a single contractor.

Having numerous, small contracts also compounded the problems of effective control and maintenance of contract documents. The sheer volume of WPPSS' paper load was brought to the public's attention in 1981 when a citizen of Vancouver, Washington, asked for copies of construction contracts, changes in contracts, cost overrun projections, payments made for cost overruns, and an equipment inventory for WNP 4 and 5. WPPSS officials informed him that he could have just what he wanted—for a minimum of $100,000. "The documents you have requested are extremely lengthy," a WPPSS official wrote. "For instance, there are 464 contracts for WNP 4 and 5. These contracts average approximately 900 pages each and involve copying 417,600 pages. The charge to you for the contracts alone would be $41,760 plus postage. The estimated costs for the contracts plus the other related documents would exceed $100,000."

Numerous small contracts also compounded the problems of negotiating labor agreements and work change orders.

Another legal restraint that added to WPPSS' cost were Washington State public bidding laws, under which WPPSS, as a joint operating agency within the State, had to operate. Before the laws were changed, WPPSS had to take the lowest bidder unless it was ready to prove in court that the firm couldn't do the job—a course WPPSS didn't choose to take. The effect of the laws on WPPSS is apparent in the following example. In 1973, WPPSS hired Bovee and Crail, a general contractor, for work on WNP 2 despite the fact that the firm was not WPPSS' top choice. WPPSS officials felt that Bovee and Crail had too little experience in nuclear plant construction. But the firm's bid of $30.9 million was far below the bids of such experienced nuclear plant builders as Peter Kiewit & Sons ($38.2 million bid) and Morrison-Knudson ($47.5 million bid). WPPSS officials doubted that Bovee and Crail, or anyone else for that matter, could do the job for $30.9 million. Their worst fears were later realized. On January 23, 1976, after a storm of charges of inexperience and cost overruns and countercharges, WPPSS kicked Bovee and Crail off the job site. By this time, the company had been paid $25.8 million. WPPSS then paid Peter Kiewit & Sons $26 million to finish the job—and had to pay out another $1.9 million to settle with Bovee and Crail's subcontractors.

(2) Inappropriate Contracting Terms

According to the WSSEUC investigation, another important factor contributing to the massive cost overruns was WPPSS' willingness to grant contractors billions of dollars through contract realignments. All early contracts were for fixed prices, but they were quickly rendered woefully underpriced by the many change orders that flowed in from the Nuclear Regulatory Commission and WPPSS management.

WPPSS mainly used three types of contracts: lump sum/fixed price; target man-hours with an incentive fee; and unit price/level of effort plus fixed fee. Each type had defects that cost WPPSS management money.

Lump/sum fixed price contracts theoretically required contractors to perform certain work for a specified fee. Large national contractors were reluctant to bid on such a basis, and the contractors who did bid generally were able to get their contracts substantially realigned after the initial award. Those contractors, according to Frank McElwee, were successful in arguing that lack of access to the work site due to congestion created by the large number of many small contractors had cost them delays and that, therefore, they required more money from WPPSS to get the job done. Exacerbating the problem was the method of financing WNP 4 and 5. Under the "capitalization of interest" financing scheme, it was in

WPPSS' best interest to get the work done and the plants operating as quickly as possible. As a result, WPPSS operated on a "fast track" construction method—a method that made it difficult, if not impossible, to hold contractors to their original bids. Fast-track construction was designed to shave years off construction time by beginning construction even while design work was still under way. Often the designs were just a step ahead of the contractors, and contractors sometimes had only a general idea of what they were bidding to build.

An example of ballooning lump sum/fixed price contracts was the electrical contract at WNP 3 and 5, which increased by almost $300 million in less than two years. In December 1979, the WPPSS Board of Directors signed a $151 million contract with the firm of Fischbach & Moore to install the electrical equipment for such facilities as the turbine building, reactor auxiliary building, and control room at WNP 3 and 5. On December 2, 1981, the Board was asked by management to approve a "realignment" of Fischbach & Moore's contract, to bring the total cost of the contract to $438 million. In papers provided to the Board of Directors, WPPSS managers said that the nearly $300 million cost overrun was due to delays caused by congestion at the job sites and extreme technical changes. Included in the $300 million realignment was $33 million worth of incentive fees that would be paid to Fischbach & Moore if electrical work was completed "on time or ahead of schedule."

Target man-hours was the second type of contracts WPPSS used. If a contracting company completed the work on time and within the "target" number of man-hours, it would be rewarded with a minimum 3 percent incentive payment. If the work was completed under the target man-hours, the incentive payment rose to a maximum rate of 15 percent. Experience on the WPPSS projects showed that target man-hour incentives alone did not significantly improve performance and instead may have greatly expanded administrative support requirements. Labor incentives had to be coordinated with the delivery of equipment and material before significant improvements in productivity could occur. Thus, the problem with the concept of target man-hours, WPPSS learned, was scheduling. Frequently, contractors would assign a job but would not have necessary materials or engineering support at the site when workers arrived.

At times there would be hundreds of workers standing around a job site on full pay for hours, sometimes days, waiting for a job to be completed or equipment and/or materials to show up. One small, local painting company, Long Painting, contracted to do fireproofing work at WNP 1 and 4, and immediately ran into scheduling problems. When work finally started, 6 months behind schedule, Long's paint crews were hampered by interference from other craft workers. Temperatures were in the freezing 30°s instead of the summery 70°s, when the fireproofing material is supposed to be applied. WPPSS had to provide heat to the building. It was still too cold, and Long's workers had to use trowels to apply the material instead of spray it on. Tiny, Long Painting suffered a $56,000 cost overrun on its $212,000 contract.

The third type of contract used by WPPSS was "unit price/level of effort plus fixed fee" contracts. They were much like the original lump sum/fixed price agreements, but they included a provision that WPPSS had the "unquestioned right to control and direct the contractors' work." This provision was added in hopes that WPPSS would be able to gain some control over costs, productivity, and quality. In the end, however, the provision only caused additional cost overruns, because contractors were then able to blame WPPSS for their failures to complete work on time within budget.

WPPSS eventually recognized the problems inherent in this type of contract. WPPSS' own administrative auditor in connection with one contract that was realigned to quintriple the size of the contractor's original fee, commented:

> Contractors obviously appreciate the removal of risks from their operations. Removal of such risks, however, permits a contractor the option of staffing his work with mediocre talent, leaving his more talented staff available for assignment to projects where high levels of expertise and ingenuity can be applied to risk taking and the rewards that go with such ventures.

(3) Failure to Hold Engineers and Contractors Accountable

The WSSEUC study found that WPPSS had failed to "get tough" with its architect-engineers and other contractors to hold down costs. As late as 1977, WPPSS had a one-man audit function that was so weak that the auditor position didn't even appear on WPPSS' organizational chart. The first auditor on the nuclear electric projects, between 1974 and 1977, was a young man who had only recently graduated from college and who had no auditing degrees or previous experience. He would turn his audit reports over to Neil O. Strand, then managing director of WPPSS. For some reason, Strand never showed the reports to the Board of Directors—but the Board never asked to see them, either.

In October 1977, Jerry Wentz, an accountant with 23 years' experience working in the California Department of Audits, became WPPSS auditor. In early 1978 Wentz began filing quarterly summaries of his department's audit activities with the Board of Directors. By then he and his staff had uncovered $20 to $25 million in questionable charges by contractors. Even then, the Board of Directors didn't do much about it or call it to their contractors' attention. For example, by late 1981, WPPSS auditors had found that Ebasco Services, Incorporated, the architect-engineer for WNP 3 and 5, had run up more than $1.8 million during 1979 and 1980 in what the auditors termed "nonreimbursible costs." Those costs, which WPPSS had already paid, included computer services provided at a mark-up of 90 percent and other expenses charged to WPPSS for executive hunting trips and conferences. Nevertheless, in September 1981, WPPSS negotiators, with that audit information in hand, had realigned Ebasco's existing contract to provide more money for overhead costs. Under the old contract, Ebasco was receiving 66.5 percent of salary for overhead for each of its permanent employees assigned to WPPSS projects. The new contract raised the overhead Ebasco received from 65.5 percent of salaries of permanent employees assigned to the projects to 70 percent. Even better still for Ebasco, the realigned contract prohibited WPPSS from auditing Ebasco overhead costs in the future. In addition, the new contract ended the old cost-plus arrangement, under which Ebasco received costs plus the fixed overhead rate plus a guaranteed 12.5 percent profit no matter how much money the contractor spent to get the job done. That was replaced by the new overhead rate, plus a fixed profit amount with incentives. This new "incentive" contract would guarantee Ebasco $45 million in profits. WPPSS claimed that the contract realignment would bring WNP 3 to completion a year early, thus saving it up to $800 million. In regard to the disputed, but already paid, allegedly nonreimbursable costs, WPPSS decided to forget about them as part of the new contract agreement. WPPSS' management felt this gesture would help motivate Ebasco to speed construction of WNP 3 and 5. Afterward, WPPSS felt that it had indeed "gotten tough" with Ebasco.

(4) Failures of Construction Site Labor Relations

Initially, WPPSS' approach to construction site activities was to contract with the architect-engineer to perform the on-site construction management role and to contract with various contractors to perform the construction work on a lump-sum basis. This approach relieved WPPSS of direct involvement in the management of on-site labor and material resources. Experience shows that this approach may have contributed up to $1 billion in labor cost overruns.

WPPSS decided early on to leave labor relations to its dozens of individual contractors on the theory that they would negotiate in their own best interests and thus WPPSS' best interests. The separate and often uncoordinated labor contracts led to numerous strikes against individual contractors that shut down entire construction sites. WPPSS' policy of not participating in construction site labor relations appears to have its origin in their earlier policy of negotiating lump-sum construction contracts. Under lump-sum contracts, the contractor usually resisted any measures that restricted their prerogatives to manage their own resources. Most of the individual contractors agreed to pay prevailing wage rates to their laborers. As a result, WPPSS found itself in a situation in which labor relations, according to Ed Hearne, Bechtel Corporation's senior labor relations representative, "was strictly a catch-as-catch-can affair." Said Hearne: "Every contractor did their own thing. In many cases, you'd have 40 different contractors on a job site, often making work assignments that were contrary to one another." Such assignments usually sparked jurisdictional disputes involving two or more unions, each claiming its right to the assignment. Sometimes, contractors were ignorant of which union had jurisdiction over certain work. Other times, a contractor would simply ignore jurisdictional rights, precipitating a walk-off by members of the "injured" union.

Most national architect-engineering firms and contractors are signatory to national trade agreements. Thus, they are obligated to have formal labor agreements for the projects they accept. Theodore Barry & Associates, in its 1979 WPPSS management study, stated that, generally, labor costs are the highest at, and work stoppages have a greater impact on, those sites that adopt the national trade agreements. The reason, said TBA, was that labor agreements are typically negotiated under the assumption that the workers will have only temporary employment with a single contractor; thus, the base wage rates and fringe benefits payments are relatively high. At the same time, each trade union has discrete national and local contracts; thus the work stoppage impact is often greater because it is difficult to coordinate the various contract provisions at each individual work site and because of the different perceptions of the local laborers and management.

However, at a site such as WPPSS' five nuclear projects, where work could continue for as long as 10 years without interruption, lower than prevailing wage rates could ordinarily be negotiated in exchange for the relative security that such a project provides. This is because major construction projects have characteristics that are somewhat different from those upon which normal labor contract negotiating assumptions are based. Had WPPSS bothered to take the initiative earlier and negotiated a site-wide agreement, it might have reduced costs and avoided disagreements that led to strikes at least in the later years of the project. As it turned out, some of WPPSS' contractors really had no incentive to settle labor disputes when they occurred at least after the third-generation contracts were put into use. These unit price/fixed fee contracts provided that total fixed fees for overhead and profit would be paid to contractors in equal monthly installments over the contract life—whether work was continuing or not. Because contractors continued to receive full payment of profit and overhead during a labor dispute, the WSSEUC investigation of WPPSS' cost overruns noted, "a contractor's incentive to bargain in good faith and reach an early settlement was reduced." When a dispute was finally settled, the contractors were often able to renegotiate for contract realignment that provided them with further incentive and progress payments because of the time lost to the strike.

As WPPSS became caught in the middle of labor disputes that were beginning to add as much as $1 billion in cost overruns, officials did decide to change their labor-management strategy—but this didn't happen until WPPSS had experienced a four-month-long strike at WNP 1, 2, and 4 in the summer of 1980. According to WPPSS' own estimates, the 22-week work stoppage added $707 million to those three plants. This experience finally showed WPPSS management that its "hands off" approach to labor relations wouldn't work, and entered into discussion with its contractors and labor leaders which led to labor stabilization agreements at all five sites.

(5) The Decision to Integrate Construction Management

Initially, WPPSS management of its nuclear construction program was based on a dual leadership role called "integrated management." In effect, WPPSS shared the responsibility for construction with its architect-engineers. By the end of 1977, WPPSS had created an organization that virtually duplicated the architect-engineers at each nuclear project site. The architect-engineers complained of "excessive oversight" and massive confusion about who was in control and who was to be held responsible for design and construction management. It wasn't until 1980, when WPPSS appointed Robert Ferguson managing director, that the duplicative organization approach was changed. Ferguson took steps to give the architect-engineers more responsibility and to keep WPPSS from interfering with their work. At the WNP 3 and 5 site, Ferguson demanded a new set of site managers from the architect-engineer firm, Ebasco, and got them. But the damage caused by the integrated management approach had already cost WPPSS. The original, 1976 estimate for Ebasco's work there was $137.5 million. By 1981, after integrated management, that amount had risen to $585 million. A Morrison-Knudson spokesman, speaking to the WSSEUC, said that the integrated management system created a situation in which "we have everybody responsible but no one was accountable for failure to perform. I think when we have everybody responsible, nobody's responsible."

(6) Failure to Delegate Authority and Responsibility

The WSSEUC study noted that because of WPPSS' preoccupation with oversight, much time and effort that could have been spent building the plants was spent justifying decisions and attending to paperwork. According to the Senate committee, critical decision-making bogged down, aggressive direction was missing, management actions were untimely, excessive central control from WPPSS headquarters emerged and was itself untimely and piecemeal, contractor control suffered lapses, a choking array of paperwork, procedures, and bureacracy grew up, and corrective actions themselves soon sank into the morass.

(7) Failure to Develop Effective Change-Order System

On construction projects like WPPSS' five nuclear power plants, an efficient system of rewriting contracts was essential to their timely completion. The WPPSS system was both cumbersome and time consuming, the WSSEUC found. "WPPSS does not have [in late 1980], nor has it ever had, an effective change order management system," the WSSEUC's report read. "The failure of WPPSS to institute such a system is a direct and principal cause of project schedule delays and cost overruns."

For example, on WNP 2 at Hanford, the plant nearest to completion, some 22,000 changes to the project's original design were made after construction started in May 1973. WPPSS estimated that 20 to 25 percent of the change orders were a direct result of changes in U.S. Nuclear Regulatory Commission standards for nuclear plant design based on the experience at Three Mile Island. But those changes do not tell the whole story. In an interview by John Hayes of *The Oregonian* newspaper, Chet Organ, assistant director of engineering on WNP 1 and 4, described the problem as a "ripple effect." He used as an example of a change order issued by NRC for increased safety tolerances for an injection pump listed on a plant's blueprint.

> Let's say that the NRC wanted to increase the margin of safety so it takes a bigger pump. The first thing one had to do was order a confirmatory analysis on the specifications to see that the change didn't violate other standards. That happened all the time with change orders.
>
> After a wait of several days, it may turn out that the floor of the nuclear reactor building must be strengthened to accommodate the additional weight of the larger pump. To strengthen the floor, the design engineers may have to double the number of reinforcing steel bars that go into the floor when it was constructed. So far, so good. Little time was lost and the change order only added a modest amount to the cost of the plant. But this was only the beginning. Months later, after most of the construction in the building was finished and the work area was crammed with equipment, wiring, huge pipes, and fittings, work crews arrive to install hundreds of pipe supports and seismic restraints that will stabilize the pipes in case of earthquake. The restraints and supports consist of bars that are bolted to steel plates that are, in turn, fixed to the concrete floor with large bolts.

One of the cardinal rules when fastening materials to concrete is that a hole may not be drilled through a steel reinforcing bar embedded in the concrete. Thus, an unforeseen problem. When the workers begin drilling the bolt holes for the restraints and pipe hanger plates, they cannot find a single place where the bolt holes can be drilled cleanly without striking a reinforcing bar. The reason: The design engineers, many months before, doubled the number of steel reinforcing bars in the concrete, making no allowance for the plates.

This kind of problem was a daily occurrence, said Organ. In this instance, a new engineering study would have to be made of both the floor specifications and the location of the seismic restraints and the pipe hanger plates. Work crews might stand idle, on full pay, for weeks while WPPSS engineers calculated a way out of the mess, one that both solved the problem and did not violate NRC safety standards.

The ripples caused by installation of the larger pump, said Organ, would continue almost indefinitely. For example, the larger pump would require a much larger electric power line to make it operate. But by the time the pump change is made, the access holes in the containment wall for the electrical conduit might already have been made. Now, workmen find that the access holes are too small to accept the larger conduit serving the bigger pump. When the time comes to drill a bigger hole in the containment wall to accommodate the larger conduit, NRC inspectors demand new engineering studies to make certain that drilling the larger hole will not threaten the safety of the wall. "By the time we get that far, most of it is no longer identifiable as being caused by the regulatory change in the beginning," said Organ.

The WSSEUC investigation revealed that in the case of one WNP project, an average of more than eight months was needed to process each change order through the WPPSS system. In some cases, change orders took as long as 17½ months to wend their way through WPPSS' complicated change-order procedure. "When the processing time for approval is excessive, the contractor in question stops or slows down work until they are instructed to proceed," the WSSEUC report noted. Life in the megabucks world of nuclear electric plant construction, WPPSS was discovering, was nothing like the relative simple task of building hydroelectric dams.

9

The Lure of Tax-Exempt Bonds

As a public agency, Washington Power Public Supply System was restricted in its source of funds. Unlike an investor-owned utility, which could obtain funds through stock issues, WPPSS was limited to the debt market, to long-term municipal revenue bonds that were to be paid off from revenues from the utility properties it operated.

In the beginning, it was estimated that the five nuclear units could be built for just a little more than $4 billion. As was the case for earlier WPPSS construction projects, financing would be through sale of tax-exempt municipal revenue bonds. Payment of interest would not start until two years after the bonds were issued. Later, when the plants were finished and operating, the bonds would be retired with revenues from the sale of electricity. Because the Bonneville Power Administration (BPA), had acquired 100 percent of the electrical output of WNP 1 and 2, and 70 percent of WNP 3, and those three plants were in effect, backed and guaranteed by a Federal agency. As a result of this WPPSS/BPA connection (plus the fact that WPPSS-generated electricity would be sold throughout the Northwest, thus spreading around the cost of the plants) bonds for WNP 1, 2, and 3, when first issued, were given a top rating of Aaa by Moody's Investors Services and AAA by Standard & Poor's Corporation. WNP 4 and 5 were backed by only their 88 public utility district and municipal utility sponsors—but there was widespread belief among the 88 Participants, investment analysts, investors, and others that in the unlikely event that the project got into financial difficulty, BPA would somehow come to the Participants' rescue.

If the "build now, pay later" scheme seemed good for everybody, it was because WPPSS' construction plan was founded on several assumptions that seemed reasonable enough at the time. One assumption was that demand for electricity in the Pacific Northwest would continue to grow at about the same, strong rate, so there would be a market for power from the five plants when they were

60845

$5000

UNITED STATES OF AMERICA

STATE OF WASHINGTON

Washington Public Power Supply System

NUCLEAR PROJECT NO. 1 REVENUE BOND, SERIES 1981D

WASHINGTON PUBLIC POWER SUPPLY SYSTEM, a municipal corporation of the State of Washington (hereinafter called the "System"), for value received, hereby promises to pay to the bearer, or, if this bond be registered as to principal, to the registered owner hereof,

ON THE FIRST DAY OF JULY,

2017

, the principal sum of

Five Thousand Dollars ($5000)

FIFTEEN PER CENTUM (15%)

WASHINGTON PUBLIC POWER SUPPLY SYSTEM

By

PRESIDENT

WASHINGTON PUBLIC POWER SUPPLY SYSTEM CORPORATE SEAL 1957 WASHINGTON

Typical revenue bond issued by WPPSS for WNP 1.

finished. Another assumption was that the plants would be finished on time and within budget. In the end, neither assumption came even close to the mark. But in the early 1970s, when WPPSS was about to begin issuing tax-exempt municipal revenue bonds for construction funds, the Pacific Northwest's need for power and WPPSS' construction management record on the Packwood hydro and Hanford steam-utilization projects were good enough reasons for investment analysts to recommend WPPSS revenue bonds.

WPPSS' Authority to Sell Tax-Exempt Municipal Revenue Bonds

In the 1950s and early 1960s, most of the bonds being issued were general obligation bonds. Traditionally issued to build schools, streets, and other public works, general obligation bonds are backed by the issuer's pledge to tax citizens as much as necessary to pay back the debt. But general obligation bonds require voter approval before they can be sold, whereas revenue bonds do not. As the number of investors interested in a tax shelter increased, revenue bonds quickly grew in popularity. Cities and states liked revenue bonds because they did not burden their credit capacity and did not require voter approval. Underwriters liked revenue bonds because they were able to charge higher sales commissions. And investors like revenue bonds because they were exempt from Federal income tax and, in the end, generally offered higher yields than fixed-income securities.

Financing of the five nuclear electric plants through the sale of revenue bonds was more than just convenient and financially rewarding. It was pretty much a necessity because Washington State law limits the authority of the state and local governments to borrow money. These limitations are designed to prohibit impulsive and misdirected overborrowing; their objective is to protect taxpayers, bondholders, and the credit of governments alike. In an effort to circumvent "pay-as-you-go" financing of many basic services, Washington State (and other state governments, as well) had created special districts and authorities, in Washington called "joint operating agencies." The most important method of financing used by these specially created districts and authorities since the 1960s has been the issuance of revenue bonds. Since revenue bonds are not paid off from the general tax revenues of a local or state government, they are not subject to constitutional or statutory debt limitations. In 1931 only 15 states permitted local governments to use these so-called nonguaranteed bonds; by 1936 the number had risen to 40, and now nonguaranteed bonds are used in every state.

In total sales, public electric power bonds account for the largest share of the revenue bond market. Between 1974 and 1978, sales of public electric power municipal bonds quadrupled, from $1.5 billion to $6 billion. By 1980, public electric power municipal bonds represented more than 13 percent of the long-term, tax-exempt market. In 1982, $87.6 billion worth of new short- and long-term revenue bonds was issued, up 543 percent from the $47.6 billion issued in 1973, and up 327 percent from the $15.6 billion issued in 1963. The key feature of municipal bonds, and the one that accounts for their substantial growth since the early 1960s, is their tax-exempt status as written in section 103D of the Internal Revenue Code. Not only are holders of municipal revenue bonds not required to pay Federal income tax on their "coupon" (interest) income, but their interest is also frequently exempt from most state and local taxes. With high rates of inflation in the 1970s pushing middle-income earners into higher tax brackets, it became ever more attractive to invest in tax-exempt revenue municipal bonds whose yields, while lower in absolute terms, usually were higher on a net after-tax basis.

Bracket Creep

A married couple earning a combined annual income of $50,000 is almost into the 40 percent tax bracket. If the couple buys a taxable fixed-income security yielding 10.75 percent, or opens a money market bank account at 8 percent, they give close to 40 percent of their earnings to the Internal Revenue Service. If they buy a long-term municipal tax-exempt bond yielding 8.7 percent, they pay only state income tax (and not even that if they live in the same state as the issuer). One can figure out when tax-exempt municipal revenue bonds might be a good investment by knowing one's tax bracket and the yield of the bonds. For example, if the middle-income couple in the 40 percent bracket purchases a municipal bond yielding 8.7 percent, their tax-exempt yield is equal to 14.5 percent. (Subtract one's tax bracket—40 percent—from 100 percent and divide the result, 60 percent, into the tax-exempt yield—8.7 percent.) This means that a taxable investment must pay more than 14.5 percent to give the couple a higher return than they would receive from the tax-exempt bonds. This arithmetic has brought huge numbers of middle-class investors into the municipal revenue bond market for the first time. In 1982, individual investors accounted for over 75 percent of the $61 billion of net new purchases. Today individual investors own about $170 billion face value of outstanding municipal tax-exempt revenue bonds, compared with only $87.6 billion in 1982.

Default Always a Possibility

Although tax-exempt revenue bonds have become popular among typically conservative investors, they are not risk free. There are worse things than paying taxes. Default, or bankruptcy, by municipalities, states, cities, and public authorities is always a possibility. Investors have suffered through more than 6,200 U.S. municipal defaults. The first one occurred in 1839 when Mobile, Alabama, failed to pay interest on $513,000 in bonds. The city could blame two major fires, the Panic of 1837, and a yellow fever epidemic. Although the depression years of the 1930s accounted for the vast majority of these municipal debt defaults (4,770), the record since the Great Depression has not been spotless even with the safeguards built in by the "New Deal" legislation of the Roosevelt Administration. There were 79 defaults in the 1940s, during the boom production of the war years. There were 112 defaults during the 1950s, and nearly 300 in the economic expansion years of the 1960s.

Defaults continued during the 1970s. Among the more well-known troubles was New York City's fiscal crisis of 1974–1975, when it failed to pay principal on $2.4 billion of short-term notes. Technically, the city never defaulted, and the noteholders were eventually paid. In another big-city crisis, Cleveland defaulted on about $15 million in notes in 1978. The city eventually refinanced the debt, stretching repayment from 1 year to 14 years. The new debt carried a higher interest rate, however. This is the way it has been for the vast majority of the defaults since 1839; bondholders eventually received their principal, though not necessarily all the interest due—but they sometimes had to wait for years—and those waiting the longest were the holders of revenue bonds.

Joint Ownership Power Systems

The general WPPSS arrangement—a joint operating agency composed of individual utilities that banded together and issued revenue bonds to construct electricity generating facilities—is not unusual. In addition to Washington, such states as Colorado, Georgia, Louisiana, Massachusetts, Texas, Wyoming, Utah, and a number of others have joint entities to construct power generating facilities financed by the issuance of revenue bonds backed by long-term power supply contracts with each of the participating municipalities.

However, power supply systems have always presented difficulties for bond analysts and investors. First and foremost is the question, whose credit or security is being pledged for the new enterprise? Are the participants jointly and severally liable for all costs and expenses, or is each participant liable only to the degree to which it is a participant in a cooperative project? If one of the partners in the joint venture is unable to pay its share, how certain is it that the other partners will assume that extra burden or will be able to find a replacement participant in time to prevent default or delay on bond interest payments? Because of some of these uncertainties in joint financings, the security for the bonds has almost of necessity been contracts for long-term power sales. These contracts must last for the life of the bonds. Such contracts have frequently been between the state authority and the power supplier, regardless of whether the power ultimately comes from a private, municipal, state, or Federal source. The power is sold only under "hell or high water" contracts that guarantee payment whether or not the power is used, whether or not the project is built, and whether or not the facility ever operates. WPPSS had such agreements with its 88 Participants.

Other Legal If's

When a new agency such as WPPSS issues bonds for construction of public power facilities, the credit analysis is usually far more extensive than that for most ongoing issuers. There are extra legal questions about the issuer's autonomy, its right to issue bonds, and its right to have a monopoly on providing power to residents within its geographic boundary. Another concern is the quality and experience of the management. In constructing free-standing nuclear electric plants there are obvious risks, including failure to complete the projects, delays that sometimes last years, and problems in engineering that can result in an improperly, inadequately, or nonfunctioning power system. Until a system is finally complete and its operations are certified as acceptable, it does not earn money. The longer the delays, the higher the costs.

Underwriting Municipal Revenue Bonds

The simplest means of selling municipal bonds is for an issuer to ask a local bank to structure a loan for a specific purpose (e.g., to purchase a piece of equipment) in the form of tax-exempt bonds, which are often sold in the community. While such sales remain numerous, the vast majority of revenue bond sales for public power, particularly when construction of large-scale electric generating facilities is involved, now come from transactions between joint operating agencies such as WPPSS and bond dealers. These bond sales are transacted in the primary, or new issue, market via a system known as underwriting and syndication. Underwriting is a common practice whereby investment bankers and commercial banks join in a syndicate contract with a bond issuer such as WPPSS to buy the entire issue at a set price, arrived at competitively or through negotiation, and then to resell that issue in smaller pieces to the general public.

The size and composition of the underwriting syndicates that purchase revenue bonds for resale vary from issue to issue. Generally, the underwriting of massive blocks of bonds cannot be handled by just one dealer or bank. This is not only because of capital limitations, but also because a bulk buyer of bonds needs to diversify its risk. Marketing syndicates usually are composed of small municipal dealer firms, larger securities brokerage and investment banking firms, and commercial banks.

Revenue bonds for public power projects may be offered through competitive bidding, but they tend to come to market via negotiated transactions with an underwriter. If the issuer decides to do a negotiated underwriting, as WPPSS did, it asks a number of underwriters to present their credentials. No bids are submitted; rather, representatives of the various underwriting syndicates meet with the issuer in advance of the offering and negotiate a coupon rate on the bonds in light of current market conditions. The issuer then selects the syndicate it believes, for a combination of reasons (fee charged to the issuer, capital strength, distributing ability, management record, municipal finance experience and expertise) will be the most competent and will market the issue at the lowest net cost. The selected syndicate assists in organizing the issue and preparing the offering circular, the official statement that describes the bonds and the issuer's financial condition, which will be made available to the public before the offering. In addition, syndicate managers conduct a pre-market sales campaign and, finally, negotiate in good faith the terms of the sale.

Early on, WPPSS hired four New York City based brokerage houses to serve as underwriters: Salomon Brothers, Goldman-Sachs, Smith Barney, and Merrill Lynch, Pierce, Fenner & Smith. The amount of money to be earned in fees and commissions through WPPSS' bonds sales was huge. For example, on WPPSS' sale of $850 million in bonds in February 1982, at 14.85 percent interest, the four brokerage houses earned $23 million in fees and commissions. Merrill Lynch, which managed WPPSS' $775 million bond sale in September 1981, earned a $5 million fee, making it the most profitable bond issue in its history.

Legal Opinion

Before selling the tax-exempt bonds, the issuer employs attorneys who specialize in municipal borrowing to analyze the issue's authorization and to give an opinion on the bonds' legality. Although these "bond counsels" are paid from the proceeds of the bond issue, they do not usually act as regular counsels to any of the participants in the issuance. Instead, they act as independent legal auditors and thereby serve two important purposes: they give an opinion on the legal validity of the bond issue as a binding obligation on the issuer in accordance with state and local laws, and they give an option on the tax-exempt nature of the interest on the issue in compliance with Federal, state, and local tax laws.

In general, the bond counsel does not address the creditworthiness of the issuer or the associated interests and market risks of the security. The bond counsel relies on the documents presented by the issuer; nevertheless, investors rely heavily on the bond counsel's opinion. In recent years, bond counsels, aware of the growing investor reliance on their opinions, have extended their duties to ensure that neither material omissions nor misrepresentations of material facts exist in the offering circular. Other duties include:

Representation of the issuer in connection with the negotiation of the bond purchase agreement with the underwriters.

Direct involvement with those parts of the official circular relating to the ordinance and the bonds. (Since many new issuers lack sufficient staff to develop an offering circular, the bond counsel frequently drafts the entire statement.)

In financings that involve power sales contracts, agreements, and other participants, preparation of those documents.

Bond counsels have departed from their traditional role of "legal auditor" to become active participants in the drafting and developing of legal standards that are applied to municipal financing. Underwriters seek "supplemental" opinions from bond counsels in other areas such as:

Power and authority of the issuers to operate as described in the offering circular.

Accurate summary of the ordinance and bonds in the offering circular.

Statement of any pending litigation against the issuer.

Assurance that the offering circular does not contain misrepresentations and omissions.

Due authorization, execution, and delivery of the bond purchase agreement by the issuer.

Since the duties and responsibilities of bond counsels have become complex, it is hard to determine to whom the bond counsel is responsible—the underwriter, the investor, or the issuer. Opinions vary from counsel to counsel. It appears that the bond counsel tries to help an issuer achieve socially useful goals and at the same time present a security worthwhile for investment to the general public. Thus the bond counsel seeks to protect the underwriter, the investor, and the issuer.

The Role of the Underwriter and Its Counsel

An underwriter of municipal revenue bonds has a dual role: representing the interests of both its clients, the issuer and the investor. In a negotiated sale, an underwriter, like a bond counsel, may be involved in structuring the financing for the security. In its responsibility to the investor, the underwriter negotiates contractual agreements with the issuer and reviews the financing and security agreements set forth in the documents supporting the offering. By doing so, the underwriter attempts to give the investor as much security as possible in terms of payment of interest and principal and debt service provisions.

Underwriters generally do not release an offering until adequate disclosure and legal verifications are completed. Although bond rating agencies may review the issue, the underwriter also should thoroughly analyze the issue, the issuer, and the financing supporting the issue.

The underwriter's counsel reviews the exempt status of the security. Relying on the opinion of the issuer's counsel, the underwriter's counsel determines whether the bond purchase agreement between the issuer and the underwriter is "legal, valid, and binding" and provides assurances that the offering circular doesn't omit any important information or contain misstatements of material facts. In addition, the underwriter's counsel may perform the following duties:

Prepare the bond purchase agreement and represent the underwriter in connection with the negotiation of that agreement.

Review the overall legal environment of the transaction with a view to ensuring that the "due diligence obligation" of the underwriters is discharged. This role involves such matters as reviewing provisions of local law concerning the authority of the issuer to conduct its operations and to issue the bond, the tax aspects of the transaction, environmental matters, construction contracts, litigation, pending legislation, and compliance with laws.

Rating Municipal Revenue Bonds

Over the last 40 years, the rating of municipal revenue bonds has grown considerably in importance and is a controlling factor in whether an investor or institution will buy a particular municipal bond. The concept of rating is simple. The rating is based on the probability of timely repayment of principal and interest. Independent rating agencies help investors by "grading" each bond issue by means of an easily recognized set of symbols, for example, A, B, or C. As the number of municipal bond issues has increased, rating of bonds by agencies such as Standard & Poor's Corporation and Moody's Investors Services, Inc., has increased in importance and complexity. In addition to helping investors, these ratings also influence the interest rate issuers like WPPSS must pay on bonds. Because of Federal and state regulations restrict many institutional investors, particularly banks and pension funds, to buying bonds of only a specified quality, these ratings can have a great impact on the success of a bond offering.

Standard & Poor's and Moody's rate bonds on request from the issuer, for a set fee based on time and effort required; they continue to re-evaluate and re-rate the bonds until the bonds have been redeemed. Municipal bond ratings categories are as follows:

Standard & Poor's Rating Categories for Municipal Revenue Bonds

AAA – Prime. Obligations of the highest quality, with the strongest capacity for timely payment of debt service. Debt service coverage has been and is expected to remain substantial. Stability of the pledged revenues is also exceptionally strong, due to the competitive position of the municipal enterprise or to the nature of the revenues. Basic security provisions (including rate covenant, earnings test for issuance of additional bonds, and debt service reserve requirements) are rigorous. There is evidence of superior management.

AA – High. Only slightly below prime quality bonds.

A – Good. Principal and interest payments regarded as safe. Debt service coverage good, but not exceptional. Stability of the pledged revenues could show some variations because of increased competition or economic influences on revenues. Basic security provisions, while satisfactory, are less stringent. Management performance appears adequate.

BBB – Medium. Lowest investment-grade security rating. Debt coverage only fair. Stability of the pledged revenues could show substantial variations, with the revenue flow possibly being subject to erosion over time. Basic security provisions no more than adequate. Management performance could be stronger.

BB—Lower. Some characteristics of investment-grade bonds, but investment characteristics no longer predominate. Generally a speculative, noninvestment-grade obligation.

B—Low. Investment characteristics virtually nonexistent and default could be imminent.

D-Defaults. Payments of interest and/or principal is in arrears.

Moody's Rating Categories for Municipal Revenue Bonds

Aaa. Best quality, smallest degree of investment risk, generally referred to as "gilt edge." Interest payments protected by a large or exceptionally stable margin, and principal secure.

Aa. High quality. Margin of protection may not be as large as for "Aaa" securities, or fluctuation of protective elements may be of greater amplitude, or there may be other elements present that make the long-term risks somewhat larger than for "Aaa" securities.

A. Upper medium grade. Many favorable investment attributes. Factors giving security to principal and interest adequate, but elements that suggest a susceptibility to impairment sometime in the future may be present.

Baa. Medium grade. Neither highly protected nor poorly secured. Interest payments and principal security adequate for the present, but certain protective elements may be lacking or may be characteristically unreliable over any great length of time. Lack outstanding investment characteristics; some speculative characteristics.

Ba. Have speculative elements; their future cannot be considered as well assured.

B. Lack desirable investment characteristics. Assurance of interest and principal payments over any long period of time may be small.

Caa. Poor standing. May be in default, or there may be elements of danger with respect to principal or interest.

Ca. Highly speculative. Such issues are often in default or have other marked shortcomings.

C. Lowest rated class of bonds. Extremely poor prospects of ever attaining any real investment standing.

The Bond Trustee

A bond trustee is a representative of the bondholder in a third-party beneficiary relationship with the issuer. The trustee enters into an agreement with the issuer with the intent of protecting the interests of the bondholders. One of the trustee's responsibilities is surveillance: It ensures that the bond covenants are adhered to, thus protecting the bondholders' interests. If problems arise with the issue, the trustee must follow certain courses of action as set forth in the bond covenants. If the issuer desires to alter provisions in the bond agreement, the trustee must approve the proposed alterations—to safeguard the legal standing and interests of the original bondholders. Chemical Bank of New York was WPPSS' WNP 4 and 5 bond trustee.

The Case of WPPSS

The participants in a municipal tax-exempt revenue bond offering work closely together in bringing the issue to the marketplace. In the case of WPPSS, all this effort was for naught. There were construction delays and enormous cost overruns. By the late 1970s, WPPSS had become the country's most expensive civil works project in history. WPPSS was being accused of gross mismanagement, and there was a growing perception that something had gone terribly wrong. In the summer of 1980, Robert L. Ferguson, a Federal energy official, was hired by the Board of Directors as WPPSS' new managing director with a mandate to clean up the mess. Ferguson made sweeping changes and generally won accolades for his progress, but it soon became evident that his efforts might be too late. Investors, too, were beginning to wonder if indeed there might be worse things than paying taxes.

10

Initiative 394

By mid 1981, as the estimated costs of the Washington Public Power Supply System nuclear program was soaring toward $24 billion, up from $4.1 billion ten years earlier, people in Washington State were asking what had gone wrong with their electric power supplier. With the cost overruns, even strong supporters of public power were coming to view WPPSS as an organization unresponsive to the public, indeed, an organization completely out of control. The pride of western populists, the shining example of the good of power when managed for the general welfare, had become an albatross, threatening the entire Pacific Northwest with bankruptcy.

The idea of locally owned electric generating facilities managed for the general welfare rather than for profit had a long tradition in Washington State. WPPSS and its populist supporters in the public utility districts had always viewed investor-owned utilities as predators. To defend themselves and to protect their interests, the public power leadership had established a closed-door, privately guarded organization. No one, not even the Washington State legislature, had control over WPPSS, although it had established WPPSS in the late 1950s. WPPSS' directors and management officials had come up through the ranks of the public utility districts. When Robert Ferguson was hired as managing director of WPPSS in 1980 with the mandate to turn around the problem-plagued nuclear program, he was the first "outsider" to join the organization at that level. The general public outside the PUDs knew very little about WPPSS. Why should they? In terms of energy, they had known nothing but inexpensive electricity generated by U.S. Government built hydroelectric dams on the Columbia and Snake Rivers. The Bonneville Power Administration was an integral part of that inexpensive electricity and whatever BPA, and certainly WPPSS, did to continue the tradition of low-cost electric power was all right with almost everyone in the state of Washington. As a result, WPPSS operated with virtually a free hand.

The belief that WPPSS could do no wrong went largely unchallenged throughout the state even as the estimated cost of the five nuclear plants was doubling and then redoubling. But there was one individual in Washington who had not grown up under the public power philosophy and did not believe WPPSS was infallible. He was Steve Zemke, a 35-year-old consumer activist. A 10-year resident of the Seattle area, Zemke had come to Washington as a graduate student in fisheries at the University of Washington. While Zemke studied at the university, he and his fellow students watched in amazement as the estimated total cost of WPPSS' five nuclear plants soared. "We looked at WPPSS and nobody was doing anything about it," said Zemke. "The governor wasn't acting. The legislature refused to do much. And the WPPSS people, of course, refused to admit they had made so much as a single mistake."

The campaign against WPPSS did not originate solely from frustrations with WPPSS' cost overruns. Initially, Zemke and his friends, who formed the "Don't Bankrupt Washington Committee" to get an initiative on the November 1981 ballot, wanted the public to take a fundamental yes or no vote on nuclear power. Feeling that such a black and white measure would not gain a great deal of public interest, the committee decided to go after nuclear power in Washington through its Achilles heel—all the money that WPPSS would later take from the ratepayers' pockets. Maybe if WPPSS had to obtain public approval before it could sell additional revenue bonds, Zemke thought, the people would have at long last some control over the single-minded supporters of nuclear power. Zemke wanted to set up a mechanism that required citizen review and approval of proposed financing for major public energy projects. Their most difficult task, the committee felt, was convincing the people of Washington State that a citizen review of WPPSS' building and financing plans was needed, justified, and long overdue. To his surprise, however, Zemke found that the electricity consumers of Washington welcomed the prospect of being able to cast a vote on WPPSS. He and his supporters obtained more than 186,000 signatures, a sizable margin over the 138,472 required to get the initiative, known as Initiative 394, on the November 1981 ballot. The initiative did not mention WPPSS; nor was it linked in any way to WPPSS. But the issue clearly was WPPSS. Zemke found that "there was a lot of anger about WPPSS throughout Washington State. If you just mentioned the word WPPSS, people would sign right up." For the first time since the nuclear projects were started in the early 1970s, the people of Washington were questioning the unlimited right of WPPSS to spend money freely, totally independent and not accountable to anyone anywhere for their actions.

Initiative 394

Some people thought Initiative 394 would save the state from bankruptcy. Others thought it would bankrupt the state. Still others thought it might do both, or neither. Whatever, passage of Initiative 394 would essentially require three things. First, it would require voter approval of the amount of money public agencies wished to borrow in order to acquire and build "major" publicly owned power plants (that is, plants producing more than 250,000 kilowatts per day). Borrowing for projects already under construction, such as WPPSS' five nuclear plants, would be subject to voter approval if their official construction cost estimates were more than 200 percent of the original estimate. Since the money would be raised by floating bonds, voting would be held within the service area of the public utility proposing a project or within the membership area of a utility group, for instance, WPPSS. Statewide voting would be held for projects proposed by state agencies.

In any balloting, voters would be asked to approve the sale of bonds necessary to finance cost overruns up to a specific limit, which could not be exceeded without another vote. Public power agencies would have to file requests for voting at least 90 days before a regular general election to get on the ballot. If there was no scheduled general election within 120 days of a filing, an applicant could request a special election "to avoid significant delay" in construction or acquisition of electric power plants.

The second thing required by Initiative 394 was that a public power agency requesting a vote would have to have an "independent consultant" study the cost-effectiveness of the proposed project compared with alternative energy sources. This "cost-effectiveness" study was also to evaluate the proposed project's reliability and availability to meet future expected energy demand on schedule. A preliminary draft of the cost-effectiveness study would have to be available for public review for at least 30 days. The final draft was to include any public comments received. The consultant would have to be approved by the State Finance Committee, which was composed of the governor, the lieutenant governor, and the state treasurer. Prior to voting, the Washington secretary of state would have to distribute a voters pamphlet providing basic information about the proposed project and allow proponents and opponents to state their views on the project. This pamphlet could be part of the regular general election voters pamphlet.

Third, Initiative 394 would set up a hierarchy of priorities for planning and financing future publicly financed energy projects. Conservation efforts would have top priority for revenue bond issues, followed in declining order by projects that would use renewable energy, waste-fuels, super-efficient fuels, and, last, nuclear energy.

Steve Zemke and other members of the Don't Bankrupt Washington Committee believed that Initiative 394 embodied legitimate goals. In fact, these goals had been lifted almost word for word from the Northwest Power Bill approved by Congress in 1980 and signed by President Jimmy Carter. Basically, the Northwest Power Bill was a public policy to be carried out by Bonneville Power Administration for the Pacific Northwest region. Zemke said that he and the Don't Bankrupt Washington Committee were not trying to run a campaign that dumped all over WPPSS. Instead, he said, "we're trying to say, what can we do to correct the present situation? And what can we do to try to head off that situation occurring again, elsewhere, in the future."

Opponents of Initiative 394

When the opponents of Initiative 394 learned that more than 186,000 citizens had signed the petition to get the initiative on the ballot, they knew they faced an uphill fight. Leading the opposition to Initiative 394 was the Western Environmental Trade Association. This group claimed that the initiative, if approved, would act like a badly aimed shotgun that could knock the financing out from under dozens of badly needed publicly built energy projects. "It will have a tremendously terrible effect on energy in general," said Richard Glaub, spokesman for the "No on 394 Committee." The No on 394 Committee retained the Los Angeles-based, nationally known public relations and advertising firm of Winner and Wagner & Associates for $100,000 to handle its anti-initiative campaign. By the time of the November election, the No on 394 Committee had spent more than $1.3 million, much of it donated by contractors and suppliers working at the five WPPSS nuclear projects. Contributions to finance the No on 394 Committee's work came from:

Intalco Aluminum $30,000
Alcoa Aluminum $30,000
J.A. Jones Const. $10,000
Zurn Industries $10,000
Reynolds Metals $30,000
Bechtel Power Corp. $25,000
Capital Development Corp. $25,000
Westinghouse Electric $15,000
Pacific Power & Light $15,000
Washington Water Power $15,000
Kaiser Engineers $32,300
Stone & Webster $10,000
NUS Corporation $500
Morrison Knudson Co. $225,000
Peter Kiewit & Sons' $25,000
Kaiser Aluminum $15,000
Chicago Bridge & Iron $10,000
Anaconda Aluminum $30,000
Ebasco Services Inc. $25,000
John Nuveen & Co. $10,000
Smith Barney, Harris Upjohn $15,000
Puget Sound Power & Light $15,000
Fluor, Incorporated $5,000
General Electric Company $25,000

In contrast, Steve Zemke's Don't Bankrupt Washington Committee raised just over $206,000. The largest single contribution to the pro-initiative 394 campaign came from Alida Dayton, a member of the wealthy Rockefeller family; she donated $30,000 to a public relations firm hired by the committee to help push the initiative. The No on 394 campaign spent an estimated $742,500 on media ads. In all 2,290 individuals and groups donated money in support of a "yes" vote, and 492 donated to the opposition.

The Press Comes Out Against Initiative 394

By and large, the press throughout Washington State was against Initiative 394, for several reasons. The *Everett Herald* editorialized that 394's main problem was that it would have the effect of managing public power utilities at the ballot box instead of in the management offices. WPPSS managers and Board of Directors, other newspapers noted, were far more experienced in building electric power plants than the general public. Certainly, there had been problems with WPPSS. But the Federal Regional Power Act had put into place some safeguards that would prevent another WPPSS building/spending spree. Most newspapers asserted that WPPSS was now under tight supervision, with much-improved management and under the watchful eye of the legislature.

In addition, the newspapers saw another drawback to Initiative 394: the cost of implementing it was largely unknown.

Editors believed that 394 would cost ratepayers more money in higher interest costs because of investor uncertainty; during the period when additional financing was needed but not yet authorized, investor confidence might be so low that bond interest rates would become prohibitively high. The opinion was that future power projects would have to be built by the investor-owned utilities, which they would not be affected by the initiaitve. The newspapers echoed what Jim Boldt, a spokesman for WPPSS, said about 394: "This initiative endangers public power; the death of public power would leave the Northwest naked against predatory aims of private power."

Many newspapers felt the fatal flaw of Initiative 394 was its basic unfairness to electricity users or ratepayers outside the State of Washington. *The Oregon Journal*, for example, criticized the initiative because the way it was worded only the people served by the public utilities in Washington that were members of WPPSS and were represented on its Board of Directors would be permitted to vote on WPPSS revenue bonds.

There is no question that the issues were complex. According to the initiative, only the voters served by the 19 public utility districts and four municipalities making up WPPSS would be allowed to vote on new bonding authority for WPPSS' five nuclear units. However, those were not the only voters ultimately responsible for the revenue bonds. WNP 4 and 5, for instance, were sponsored by a total of 88 public agencies in Washington, Idaho, Oregon, and Montana. Only 20 of the 88 were members of WPPSS, so the remaining 68 would have no say in the matter, although they would be affected by the vote. On the other hand, customers of some WPPSS members, Seattle City Light, for example, would be able to vote on WNP 4 and 5 bonds although those WPPSS members were not participating in their construction.

Critics believed Initiative 394 could be successfully challenged in court because of its obvious faults as well as its restraint of trade. Many newspapers held the initiative up as a poorly conceived effort to gain some fiscal control over WPPSS. "If it passes," wrote *The Oregon Journal*, "take it to court and let the lawsuits begin."

Overwhelming Voter Approval of Initiative 394

On November 3, 1981, Washington voters overwhelmingly approved Initiative 394—by such a wide margin that even Steve

Zemke and the other members of the Don't Bankrupt Washington Committee were stunned. In unofficial tallies, the measure was approved, statewide, 495,013 to 356,784. A much tighter race had been predicted by opponents and proponents alike. In Grays Harbor County, where WNP 3 and 5 were being constructed, voters approved the initiative by a vote of 11,023 to 3,746. In Okanogan County, a rural, ranching county which was the home of WPPSS Executive Board Chairman Nick Cain, it was approved 4,252 to 3,166.

While the pro-initiative forces were celebrating the fact that they had carried six of the seven counties that had members on the WPPSS Board of Directors, WPPSS itself was telling the media that the initiative would not stop it from continuing with its construction plans. Robert Ferguson said, "Our financial advisors in New York have told us we would just go ahead as we had originally planned." WPPSS could live with the new law, assuming it was found to be constitutional. "The initiative doesn't go into effect until July, 1982," said WPPSS spokesman Tom Hunt. "We have a year in which to sell more bonds [without voter approval] to fund the plants, and some will be marketed within 90 days." Hunt said that a court challenge would be filed immediately on the grounds that voters retroactively changed the covenants signed by WPPSS with buyers of its bonds.

Banks Sue to Overturn 394 Vote

On December 5, 1981, three banks that were serving as bond trustees for WNP 1, 2 and 3—Seattle-First National Bank, Continental Illinois National Bank of Chicago, and Morgan Guaranty Trust Company of New York—filed a complaint in the U.S. District Court of Washington, Eastern Division (Spokane) to overturn Initiative 394. (At the request of the attorney general of the State of Washington the venue was later changed to the U.S. District Court, Western Division.) Their complaint, which named the State of Washington as the defendant, claimed in part that the initiative "caused immediate, direct, substantial and irreparable injury to all bondholders by substantially diminishing bondholder security."

On April 9, 1982, the U.S. Department of Justice initiated a similar lawsuit in the U.S. District Court, Western Division of Washington, challenging Initiative 394's constitutionality on behalf of the United States of America and asserting certain rights and interests of Bonneville Power Administration. The complaint challenged the initiative as it applied to the net billed projects, alleging, in part, that the initiative impaired the existing contract relied upon by BPA and its preference customers for the construction and operation of WNP 1, 2, and 3. The U.S. Department of Justice said that the initiative also interfered with contractual arrangements approved by Congress to serve the energy needs of BPA's customers throughout the Pacific Northwest and disenfranchised the majority of BPA's ratepayers, who would have to bear the costs of amortization and interest on existing bonds. On April 29, 1982, the court stated that it would consolidate the two complaints. The trial was scheduled for June 28, 1982, and the court indicated that it would render its opinion shortly thereafter.

The bond market was badly shaken by the passage of Initiative 394, and investors begin fearing for the billions of dollars they had already lent WPPSS. Would Washington voters allow WPPSS to obtain more funds so that it could continue with construction and pay interest on bonds previously issued? If not, what might happen to the unfinished nuclear projects? Investors didn't know what course to take, and investment brokers were unable to predict how long matters might be up in the air. Regardless of what was decided about Initiative 394's legality, one thing was fairly certain: there would be construction delays—and thus higher final costs. The initiative might eventually be found unconstitutional, but it could cause problems in the meantime. If the constitutionality of 394 were upheld and voter approval for the issuance of bonds with respect to net-billed projects was not obtained, WPPSS might have to delay or terminate the projects. The investment houses believed that any such action on any one of the five projects could adversely affect all of them. The delicate financing scheme WPPSS had put together was unraveling at an uncontrollable rate.

The long-awaited decision came on June 30, 1982, when Seattle-based U.S. District Judge Jack E. Tanner ruled that Initiative 394 was unconstitutional as it applied to WNP 1, 2, and 3. (The effective date of the ruling was delayed until at least April 15, 1983, to allow time for an appeal.) Tanner declared that the initiative violated Article 1, Section 10, of the "Contracts Clause" of the U.S. Constitution, which precludes states from passing laws that impair existing contracts unless it is reasonable and necessary.

Although Tanner had ruled in favor of the banks and WPPSS, there was still a cloud over WPPSS' future. For one thing, there would be appeals. Perhaps more important, the campaign to pass Initiative 394 the previous November had left WPPSS, with its cost overruns and mismanagement, exposed for all the world to see. This was the first most investors knew about WPPSS' financial transgressions—and the impact was immediate. WPPSS' sources of construction and interest capital were shut off.

On January 12, 1983, the U.S. District Court of Appeals in San Francisco upheld Judge Tanner's 1982 ruling that Initiative 394 could not be applied to WNP 1, 2, and 3. The victory for WPPSS and the three banks didn't really matter anymore, however, for WNP 4 and 5 had by then been terminated and WPPSS was desperately trying to keep WNP 1, 2, and 3 from suffering the same fate. Zemke and his colleagues in the Don't Bankrupt Washington Committee had done what no one else had done: expose the fatal flaws in WPPSS' energy planning and construction program, which in turn led to a shutoff WPPSS' financial spigot.

11

Washington Nuclear Plants 4 and 5 Terminated

Long before WPPSS suffered its crushing no-confidence vote through passage of Initiative 394 it was becoming apparent that projections of long-range energy needs made in the late 1960s and early 1970s had been significantly overestimated. By the end of the 1970s, concern about WPPSS' large cost overruns was being voiced throughout the Pacific Northwest. The rising interest rates of the late 1970s and the delays in plant construction were increasing WPPSS' revenue requirements dramatically. Investment houses, however, did not appear to be worried about WPPSS' ever-more-frequent borrowing. Nor did WPPSS appear to be concerned. Bonneville Power Administration wasn't sure what to think, and in 1978, it hired Theodore Barry & Associates to make a management study of WPPSS. The study report, published in January 1979, criticized WPPSS for inefficient contract auditing procedures, an inept public relations program, and, most important, a management system unable to control project cost, progress, and quality.

About a year later the Washington State Senate Energy and Utilities Committee began its inquiry into WPPSS' nuclear projects, particularly the delays and cost overruns. After a year, it concluded: "Mismanagement has been the most significant cause of cost overruns and schedule delays on the WPPSS projects."

In a move to "resolve the uncertainty concerning the advisability of continuing the current construction schedule of the plants," the Washington State Legislature, 1981 session, adopted the recommendation of the Senate Inquiry Committee to authorize a "prudent review" of the status of WNP 4 and 5 by enacting Senate Substitute Bill No. 3972. The joint Washington Energy Research Center of the University of Washington and Washington State University was authorized to conduct an "independent study" of WNP 4 and 5 through the Office of Applied Energy Studies at Washington State University. The investigators were assigned specific tasks:

1. To determine the need for WNP 4 and 5;
2. To assess financing support for construction of the projects;
3. To determine costs and schedule of these projects;
4. To determine the cost of power to the ratepayer as a result of these projects;
5. To determine the average electrical power rates for participating utilities;
6. To assess the outside regional market for WPPSS power;
7. To determine the cost effectiveness of alternatives to WNP 4 and 5; and
8. To determine the cost impact of a temporary power supply deficit as compared to a temporary supply surplus.

The study director was instructed to present a report to the Washington State Legislature in the (spring) 1982 regular session.

While these studies were being done, key events affecting WPPSS' future were developing rapidly. In the spring of 1980, the WPPSS Board of Directors replaced Managing Director Neil Strand with Robert Ferguson. Ferguson moved quickly to control WPPSS' economic downslide by bringing in new executive management personnel, revising cost estimate procedures, and initiating a major engineering practices review. Then on May 29, 1981, in a totally unexpected announcement, Ferguson recommended a moratorium on WNP 4 and 5 construction to allow time for engineering revisions and construction reassessment. The Board of Directors, on June 14, 1981, approved their new managing director's recommendations—but mothballing wouldn't become a reality until it had been approved by all 88 Participants.

In the meantime, WPPSS' financial crisis was worsening. On July 23, 1981, Washington Governor John Spellman and Oregon Governor Victor Atiyeh asked Oregonian John A. Elorriaga, chairman, U.S. National Bank of Oregon and Washingtonians George H. Weyerhaeuser, president of Weyerhaeuser Corporation, and Edward E. Carlson, chairman, UAL, Inc., to "investigate the economic consequences to the region of the future construction or disposition" of WNP 4 and 5. "Two more prestigious or respected Washingtonians cannot be found for the job," said Spellman of these appointees.

Weyerhaeuser, then 55, a resident of Federal Way, Washington, was president and chief executive of Weyerhaeuser Corporation, the second largest employer in Oregon and Washington and the seventh largest U.S. exporting firm. He was also was a member of the board of directors of Boeing Company, Safeco Insurance, and Standard Oil of Indiana. Carlson, then 70, lived north of Seattle in King County and was chairman of UAL, Inc., of Chicago, parent company of United Airlines, Western Hotels, and GAB Business Services. He was also a director of Deere and Company, SeaFirst Corporation, Seattle-First National Bank, and Dart & Kraft Inc.

Spellman and Atiyeh had created the three-member panel after being approached by representatives of publicly owned utilities who feared that WNP 4 and 5 could be forced into an untimely termination. The utilities hoped that a two-state commission could, by analyzing the effects of termination, forestall a sudden abandonment of the two power projects. They pointed out that if WNP 4 and 5 were terminated the 88 utilities sponsoring them would be forced to repay $2.25 billion in debts almost immediately. Without another source of financing, the money could some only through greatly increased power rates paid by the sponsors' customers. Peter Johnson, head of the Bonneville Power Administration, expressed hope that the governors' panel would turn over new ground in its analysis, looking at factors that somehow had escaped scrutiny in earlier studies. "It will become clear," he said, "we must find a long-term solution."

Though a long-term solution might be the ultimate goal, Robert Ferguson, WPPSS' managing director, was concerned just about trying to keep work on WNP 4 and 5 going for just the next few months. Although the mothballing concept had been approved by the Board of Directors in June, work would continue until all had agreed on a plan. To give WPPSS the needed financial expertise during WNP 4 and 5 mothballing activities, Governor Spellman expanded the Board of Directors. His choices were panel members Edward E. Carlson and Charles Luce, C. Michael Berry, president of Seattle-First National Bank, and William E. Roberts, vice president and owner of Morgan Park, Inc., of Portland, Oregon.

Regionalizing WNP 4 and 5's Costs in Order to Save Them

On July 27, 1981, the WPPSS Board of Directors adopted a 1982 construction budget of $13 billion for finishing the five nuclear projects, even though funding for two of the plants remained in doubt. The total cost of the five projects was now being estimated at nearly $24 billion. WPPSS' 1982 budget was approved with only minor modifications, mainly an amendment to encompass a resolution that every effort be made to regionalize the costs of WNP 4 and 5 before all construction money ran out in October 1981. As WPPSS saw it, regionalization would require BPA to acquire the output of WNP 4 and 5, which would then be sold throughout the region, thereby spreading the burden of paying for the two plants. Peter Johnson, head of BPA, refused to commit the BPA to any course of action involving WNP 4 and 5, saying only that it would be a disservice to make any announcement before the governor's panel had finished its investigation.

With WPPSS' public announcement of its 1982 fiscal year budget, which called for more than $9 billion in debt to be issued in order to complete WNP 4 and 5 plus an additional $4 billion to complete WNP 1, 2, and 3, Washington Governor John Spellman delivered a critical blast at WPPSS and the crisis it was creating with its capitalization-of-interest financing. To complete all five nuclear plants, it was appearing, WPPSS would have to borrow an average at least $1 billion a year for the next 10 to 15 years. "It's an extreme, acute emergency and a threat to the credit of both private enterprise and governmental agencies," the governor said. "[The participants in WPPSS] are public utilities, and they are going to have to be public utilities, with the word public underlined."

WPPSS' Bond Rating Revised Downward

With the release of its 1982 fiscal year budget, which called for spending billions of dollars more to complete the nuclear plants, WPPSS increasingly was being looked on as a credit risk. On June 11, 1981, Moody's downgraded WNP 4 and 5 bonds ($2.25 billion outstanding) from A1 to Baa1, and on June 19 Standard & Poor's lowered its rating from A+ to A. (Bonds for WNP 1, 2, and 3 remained highly rated—Aaa by Moody's and AAA by Standard & Poor's—because of the Net-Billing Agreements with BPA.) John Nuveen & Company, the only national investment firm devoted exclusively to underwriting, marketing, and trading municipal bonds and tax-exempt bond funds, called the downgrading of WNP 4 and 5 revenue bonds in the summer of 1981 an "overreaction":

> [WPPSS WNP 4 and 5] are still intrinsically at least comparable to other A-rated securities and 1981 developments will turn out to strengthen, rather than weaken, the underlying credit. All of the WPPSS bonds presently outstanding continue to be credit-worthy due to Participant take-or-pay, hell-or-high-water backing.

Although municipal bond specialist John Nuveen & Company remained bullish on WPPSS bonds, other investment brokerage houses were not. Drexel Burnham Lambert, Inc., believed that a "rating suspension would have been a more appropriate response to the construction slowdown." "We consider our first responsibility to be to the bondholders, and to that end we will [suspend our ratings] on WNP 4 and 5 until such time as we are satisfied that we can evaluate the full impact of the [proposed construction moratorium]." Merrill Lynch also was not optimistic. In July 1981 Merrill Lynch released a report that concluded that bonds issued to finance all five projects could be in serious jeopardy unless the participating utilities raised electric rates immediately to begin paying debt service on WNP 4 and 5. To finance further work on WNP 4 and 5 through the municipal bond market, Merrill Lynch stated, the perceived risk had to be reduced to levels that would permit bond traders to accept WNP 4 and 5 debt as "medium grade" and sell them to a broad spectrum of the market. Merrill Lynch also warned against BPA involvement in WNP 4 and 5 because such involvement could reduce the ratings on WNP 1, 2, and 3 bonds.

Saving WNP 4 and 5

By the end of July 1981, the 88 utilities that owned WNP 4 and 5 were threatening to abandon the two unfinished nuclear plants unless they received some financial help. Seeing that its traditional Wall Street financing was being cut off, WPPSS' proposed solution to its financial problems was to regionalize the facilities. WPPSS' Participants' Review Committee said it would "pursue all available options" to solve the financial problems. Among the recommendations offered by representatives of the municipal bond market and the Participants' Review Committee (and later, by the governors' panel) were:

1. The 88 Participants and their ratepayers should share

in the risks of these projects by beginning to pay interest costs. (Originally a 50 percent assumption of interest costs was considered acceptable; the governors' panel suggested 100 percent.)

2. The financial base of the entire WPPSS program should be expanded. This might be done in several ways: BPA net-billing arrangement for WNP 4 and 5 similar to that for WNP 1, 2, and 3; partnership among WPPSS Participants, investor-owned utilities, and direct-service industrial customers; or an expanded partnership including the state governments of Washington and Oregon, the investor-owned utilities of California, and even the government of California.
3. The management of WPPSS should show it is capable of controlling costs and schedules, of completing the construction of WNP 1, 2, and 3, and of producing power at cost-effective rates.
4. The uncertainties surrounding Initiative 394 should be resolved, either by having it declared unconstitutional (it later was, on June 30, 1982) or by voter approval of the bond issues needed to complete WNP 4 and 5.

In August 1981 the Participants' Committee agreed to assume 50 percent of interest during construction, beginning on March 1, 1983. Until then the interest would be paid from a one-year reserve interest fund held by WPPSS. But Wall Street remained unconvinced about the Participants' Committee's sincerity as well as its power over the 88 Participants to make them comply with their agreement. Eileen Titmus Austin, vice president of the investment house of Drexel Burnham Lambert, Inc., questioned the agreement. She suspected the Participants' Review Committee was just "buying time" in the hope that BPA would step in and bail them out.

Many people had reason to believe that the 88 participating utilities were trying to "buy time," in the hope that WNP 4 and 5 would be made part of the Bonneville Power System and, thus, that BPA would purchase the output of these two plants should they ever be completed. WNP 4 and 5 had been the focus of debate during writing of the Pacific Northwest Power Planning and Conservation Act between 1977 and 1980. Representatives of the 88 Participants in WNP 4 and 5 had lobbied hard to have those two plants "grandfathered into" (that is, included in) the Act, but they failed because the act stressed conservation, not additional construction.

The Pacific Northwest Electric Power Planning and Conservation Act, which was signed into law on December 5, 1980, by President Jimmy Carter, was designed to manage the region's future electricity needs. The Act made BPA responsible for supplying power under the direction of a new planning organization, the Northwest Power Planning Council (NPCC). It required that all new power sources acquired by BPA be justified as the most cost-effective available and that conservation and renewable resources be considered before any central generating stations—such as WNP 4 and 5—were brought into the Federal system. Although the NPCC was sympathetic to the 88 utilities' economic plight and was concerned about general uncertainties over WNP 4 and 5, it was unable to offer them any hope of being brought into the Federal system. First the NPCC had to finish work on its 20-year forecast of the region's energy needs and an energy plan to meet those needs. The work would not be completed until the spring of 1983. Only then would NPCC be able to guide BPA's resource-acquisition program. Even if additional central generating facilities were deemed necessary, the NPCC could act with respect to specific plants only through the mechanism provided in section 6(c) of the Act. Under that provision, it was BPA that had to make the initial determination of whether a specific major resource was needed, cost-effective, and consistent with the priorities and other considerations in the Act. The NPCC then had 60 days to respond to BPA's recommendation to acquire that specific resource. If the NPCC approved or took no action, the acquisition could go forward. If the NPCC found the proposed acquisition to be inconsistent with its energy plan or with provisions of the Act, BPA would have to obtain approval from Congress before acquiring the specific resource.

In addition to questioning the sincerity of the Participants' Committee, some on Wall Street were skeptical about the 88 Participants' willingness to pay the interest on bonds, noting that they had balked at an earlier plan recommended by their own WPPSS Participants' Committee under which they would start paying 50 percent of the interest immediately (rather than in 18 months). There were other obstacles to cleaning up WNP 4 and 5's financial problems, among them a lawsuit filed by the city of Seattle and other municipalities challenging BPA's low-cost power sales contracts to big industrial customers. Another potential problem was "hell-or-high-water" provisions in the Participants' Agreements which had never been tested in court. Finally, there was WPPSS Managing Director Ferguson's proposal that WNP 4 and 5 be mothballed. Municipal bond analysts felt that mothballing would make it even more likely that WNP 4 and 5 would never be completed. Certainly, it would increase their price tag.

Throughout July 1981, the story of WPPSS' economic downslide dominated the media. Adding to WPPSS woes was the petition drive that summer to get Initiative 394 on the general election ballot in November. At the same time, the governors of Oregon and Washington were forming the special three-member panel to examine the impact of WNP 4 and 5 on the economy, utilities, and citizens of the Pacific Northwest. The day the panel was announced, BPA offered WPPSS $100 million—enough to keep WNP 4 and 5 from failing to make its mid-August interest payments to WNP 4 and 5 trustee, Chemical Bank of New York. Now WPPSS could continue with WNP 4 and 5's construction until October—and hope for a miracle so mothballing would not have to take place. After accepting BPA's donation, WPPSS announced that it would immediately have to find additional participants in WNP 4 and 5 if there was to be any hope of saving the two plants from termination. According Managing Director Ferguson, WPPSS was hoping to persuade the Pacific Northwest's investor-owned utilities and major industries to commit funds to save the nuclear units.

On September 18, 1981, the governors' special panel released its findings. They had concluded that WPPSS could no longer finance construction of WNP 4 and 5. First, the panel recommended that work on WNP 4 and 5 be halted indefinitely, instead of the 2 to 2½ years previously recommended, to allow WPPSS to find a way out of its financial crisis. Second, it urged the region's investor-owned utilities to "invest" their own money in WNP 4 and 5 in order that they might be completed as planned. The only alternative to mothballing, said the panel, was termination—and that would have disastrous financial consequences for the region.

By and large, the regional newspapers supported the slowdown of WNP 4 and 5. The *Tri-City Herald* of Richland, Washington, home of WNP 1, 2, and 4, said:

> It will hurt—the slowdown of WPPSS plants No. 4 at Hanford and No. 5 at Satsop. It will be painful for the workers who are being laid off. And for the families. There will be pain, too, for the Tri-Cities. Our economy will feel a pinch because of the lost payroll. But the slowdown of construction seems the only practical solution to a serious and complicated problem. It was a difficult decision for the utilities. Everyone shares the hope of

> WPPSS officials that the bond market will improve soon and long-term financing can be obtained at a reasonable cost. The slowdown will give everyone time to ponder the lessons taught by past events.

Most of the 88 Participants in WNP 4 and 5 supported the WPPSS Board of Directors' decision to slow down construction on WNP 4 and 5. However, they also demanded a long-term solution.

The Mothballing Plan

By mid-September 1981, WPPSS officials had hammered out a mothballing plan designed to save WNP 4 and 5. They had found that slowing down construction on WNP 4 and 5 was an easy decision; finding ways to minimize the financial risks to the 88 Participants was infinitely more difficult. The cost of putting WNP 4 and 5 on mothballs until mid-1983 was estimated at $150 million. The plan called for investor-owned utilities to contribute 35 percent, or $52.5 million. Industrial firms (mostly aluminum companies, known as direct-service industries) were called on to contribute about $37.5 million, or 25 percent of the amount needed. The remaining 40 percent, or $60 million, would have to come from the 88 Participants through their ratepayers. The aluminum companies said their support was contingent on agreement by public utilities, including Seattle, to drop a lawsuit challenging their new 20-year contracts with BPA. (In accordance with the Pacific Northwest Electric Power Planning and Conservation Act of 1980, BPA had offered new power contracts to existing direct-service industrial customers. Immediately, on August 31, 1981, a group of public utilities headed by Seattle City Light had filed a lawsuit against BPA, claiming that the new contracts provided the direct service industries with greater access to power than allowed under BPA's preference clause.) After WPPSS offered a series of compromise proposals, the 17 direct-service industries finally agreed to contribute $30 million toward mothballing, though the public utilities continued with their legal action.

By October 23, 1981, WPPSS had achieved tentative agreements with most of the 88 Participants, investor-owned utilities, the direct services industries on mothballing WNP 4 and 5. The plan ws now 30 days overdue and WPPSS did not know exactly when money might run out. Ray Foleen, a consultant working for the WPPSS

'I'm storing this suit for a couple of years and I'd like to buy $150 million worth of mothballs'

<u>Spokesman Review</u> October 25, 1981; reprinted by permission.

Participants' Review Committee, expressed disappointment that after much hard lobbying with the 88 Participants only seven of the 88 Participants had voted against the mothballing plan (four small public utility districts in Oregon and three in Washington). He believed that the Participants saw mothballing as the first step toward termination. "They're looking for a way to get out of them." To offset losses resulting from several Participants refusing to participate in the mothballing plan, WPPSS had asked the other Participants to contribute eight percent over and above their original share. Many of the Participants said they could contribute only two percent extra, and Foleen felt the entire mothballing plan might "self-destruct" if even a few utilities carried out their threat. The immediate problem facing WPPSS was that it owed $60 million by December, with the remaining $90 million to be spread over 18 months of the 24-month mothballing plan. In November WPPSS received a $60 million contribution (loan) from the larger PUDs, but continuing doubt about the long-term solvency of the plan caused other PUDs to hesitate about contributing their share. If the mothballing plan did not work out as anticipated, the PUDs feared, they would be forced to pay all of the interest due on the two projects' $2.25 billion debt between March 1982 and June 1983, when the mothballing plan would expire. WPPSS officials were hoping that by June 1983, the Northwest Power Planning Council planners would have determined that WNP 4 and 5 were feasible and needed and therefore they would have been brought into the Bonneville Power System.

On November 3, 1981, Washington voters overwhelmingly approved Initiative 394 and WPPSS Directors put off plans to sell more bonds for WNP 1, 2, and 3. By December, uncertainty over whether WPPSS would receive promised mothballing money had placed the future of WNP 4 and 5 in doubt; 27 of the 88 Participants had not made either their November or their December mothballing payments. Ray Foleen said the future of WNP 4 and 5 "is just as precarious now as it was in October" when public utility districts and others initially agreed to participate in a mothballing "rescue" effort. Nor was there any way of knowing if these 27 Participants would make any financial contributions by end of January 1982, the absolute deadline if WNP 4 and 5 were not to be terminated.

On January 5, officials of the Clark County public utility district, which owned a 10 percent share in WNP 4 and 5, changed their minds and decided not to participate in the mothballing plan. Clark County PUD's share of the mothballing plan would have amounted to $9 million, and the PUD's commissioners were not willing to pick up the extra costs of those utilities that had decided earlier not to participate. The day following the Clark County PUD's decision not to participate, Dan Evans, chairman of the Northwest Power Planning Council, pleaded with the Participants who had failed to contribute to the mothballing plan. Evans said he was supporting the mothballing plan because he was not convinced that the power from WNP 4 and 5 would be unnecessary. Ed Morris of the WPPSS Participants' Review Committee said that the Clark PUD's vote, in all likelihood, meant termination of WNP 4 and 5. The day after the Clark County PUD voted not to participate in the mothballing plan, Moody's Investor Services suspended its credit ratings of WNP 4 and 5 bonds, and also pronounced it was reviewing its ratings of WNP 1, 2, and 3 bonds. Moody's said its ratings on WNP 4 and 5 would be suspended until the projects were either mothballed or terminated.

The same day Moody's suspended ratings on WNP 4 and 5 bonds, the findings of the study commissioned by the Washington State Legislature were released by the Washington Energy Research Center. The report concluded that the region might not need WNP 4 and 5 any time before the year 2000. It also noted that it could be more expensive for ratepayers to resume construction on 4 and 5 after mothballing than to scrap the plants immediately.

With the mothballing plan rapidly falling apart, Robert Ferguson said that he'd recommend termination of WNP 4 and 5 if the mothballing arrangement couldn't be pulled together soon. He gave the 88 Participants until January 15 to decide whether the two projects would be mothballed or terminated. On January 12, the Tacoma City Council—which owned the second biggest share of WNP 4 and 5—voted to reject mothballing. That vote made termination almost certain. Thirty-three Participants out of the 88 in the end had refused to share in the mothballing costs. In spite of intensive behind-the-scenes lobbying by Washington Governor Spellman, WPPSS Managing Director Robert Ferguson recommended to the Executive Board on January 15 that WNP 4 and 5 be terminated as of January 22. At the same WPPSS Board meeting, three of the four Board members appointed by Governor Spellman, Charles E. Luce, Edward Carlson, and C. Michael Berry, resigned together. They said in a joint letter of resignation that they felt they were no longer needed on the WPPSS Board now that WNP 4 and 5 was being terminated. They had been appointed to the Board because of their financial expertise and, of course, termination

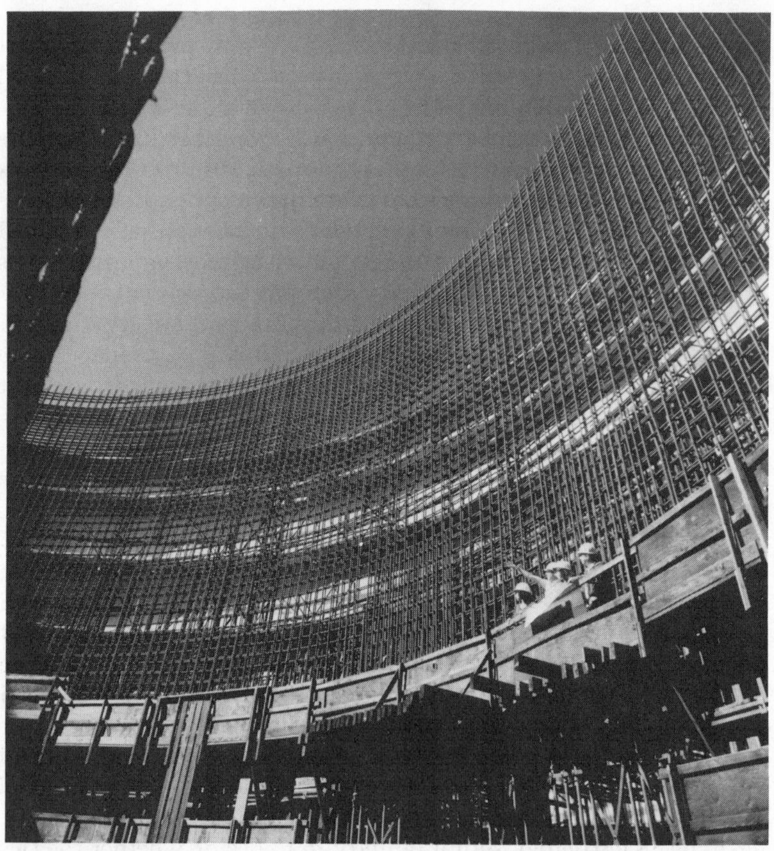

View from inside the unfinished containment building on terminated WNP 4, circa 1982.

would have required their financial expertise even more. It was believed that the three men wanted off the Board to avoid getting caught up in the expected flood of litigation after termination.

News of WPPSS' termination of WNP 4 and 5 was greeted in the financial community without worry. Halting work on WNP 4 and 5 was viewed by many analysts and investors as the only alternative to permit financing to continue on WNP 1, 2, and 3. "I'm encouraged to see a concrete plan, although I would have preferred to have seen it sooner," said Eileen Titmus Austin of Drexel Burnham and Lambert, Inc. "They are running against the clock for funds for WNP 1, 2, and 3. All the loose ends on a termination plan for WNP 4 and 5 must be tied up, in my opinion, before there can be a successful bond sale on 1, 2, and 3." Said Robert Adler, a municipal bond analyst with Preston, Ball & Turben: "I became worried that they might fool around so long on this issue [termination] that they could inadvertently set off a default."

Oregon Congressman Weaver and Wall Street Analysts

Some critics claim that Wall Street forced WPPSS into an untimely mothballing. Is there any truth to the claim? Perhaps. Oregon Congressman James Weaver, a longtime foe of WPPSS' nuclear power plants, claimed credit for Wall Street's increasing reluctance to finance WNP 4 and 5 – and those financing difficulties were the chief reason for WPPSS' Managing Director Robert Ferguson's May 29 recommendation of a slowdown of work on the two troubled plants.

In a letter to his constituents and friends dated June 18, 1981, Weaver claimed he had told Wall Street "the inside story" on the "moratorium which will save us from ruinously expensive power." Part of the "inside story," he said, was "my own efforts to let the nation's financial community know the truth about the WPPSS fiasco." Weaver said that he had met with about 80 Wall Street analysts in an informal session in late May before Ferguson's May 29 recommendation regarding WPPSS' decision to mothball WNP 4 and 5. Weaver said that he told the Wall Streeters that the projects were a "boondoggle" and that the power was not needed in the region. According to Weaver, the meeting was arranged by several analysts who had been calling him to ask about the WPPSS bonds, and several other dozen other analysts "just showed up."

Some of the Wall Street analysts were "downright hostile," said Weaver. Those were the ones who had been making money off the bond issues, he said, or those who "were stuck with some WPPSS bonds that they hadn't been able to sell to widows or orphans." Weaver wasn't sure his presentation had had any effect on the subsequent downgrading of Moody's rating of WNP 4 and 5 bonds from A1 to Baa1. "I'd like to take credit for this, but I have no way of knowing how much, if any, effect it had," he said. "But the facts about those plants speak for themselves." Weaver said he didn't consider it meddling to try to inform the financial community of his doubts about the plants. "All I'm trying to do," he said, "is to try to save the Northwest and my constituents from bankruptcy."

12

Litigation

After efforts to save WNP 4 and 5 failed, Washington Public Power Supply System turned to cutting its losses by selling off whatever it could of the two partially completed projects. There had been some talk that maybe the plants could be sold in their entirety, but no buyer could be found. There had also been some talk of California utilities taking over the plants, but that did not work out either. So WPPSS' Board of Directors decided to sell off whatever they could of WNP 4 and 5, piece by piece if necessary, thereby picking up perhaps 10 cents on the dollar of their investment. Nuclear-industry insiders knew that there would be a limited market for the equipment, even though it had never been used, because so much of what goes into a nuclear plant is custom made. It looked as if most of the enormous debt would have to be paid by the 88 Participants.

Many of the tiny public utility districts were worried that their rates might rise as much as 100 percent within the next six months. Klickitat County PUD feared the effects termination might have on its electric rates and had tried to file for bankruptcy in an effort to avoid paying its debt on WNP 4 and 5 of $3,648,000, only to be told by legal counsel that bankruptcy was impossible because it had the power to raise electric rates to obtain more revenue.

For WPPSS, termination of WNP 4 and 5 had been an immense disappointment—but now that it had happened, it did not seem to be all that bad. Officials were eager to turn their full attention and resources to completing WNP 1, 2, and 3. With the termination of WNP 4 and 5, WPPSS now had at least one year of grace before having to start paying back its $2.25 billion debt. If this debt were stretched over 30 years, interest costs of $6 billion would be added, making the total repayment $8.25 billion. WPPSS officials didn't think payback of the debt would be a serious problem for the 88 Participants. They estimated that electric rates among the Participants would have to rise only 10 to 15 percent in order to pay off all the WNP 4 and 5 bondholders.

In their mood of optimism, WPPSS officials must have been taken aback by the next turn of events. There had been some warning. For months before the termination, people in the state government and in the public utility districts had been looking for someone to blame for WNP 4 and 5, and most often their target was the Bonneville Power Administration. But after termination, the chorus grew loud, then angry. Portland General Electric Company Chairman Robert Short started off by criticizing BPA for foot-dragging in implementing the Pacific Northwest Electric Power and Conservation Act, which had taken effect in December 1980. According to Short, BPA had not moved as fast as it might, especially in acquiring new power resources. Had BPA acted more quickly, said critics, WNP 4 and 5 could have been brought into the Bonneville Power System and termination would have been unnecessary.

Short's criticism of BPA was mild compared with the complaints coming from the 88 Participants in WNP 4 and 5. The leader of the opposition was Oregon Congressman James Weaver, the man who came close to claiming credit for WNP 4 and 5's mothballing. A day before WPPSS' Board of Directors formally terminated WNP 4 and 5, Weaver stated in a letter to the editor of *The Oregonian* newspaper (January 21, 1982) that he doubted whether the 88 partners in WNP 4 and 5 were obligated to pay regardless of the plants' status. Said he:

> I do not intend to allow my constituents to be played for suckers any longer. The existing WPPSS contracts require that we pay the costs, come hell or high water. Ridiculous. Why should the bond holders, who are making 15 percent interest off billions of our dollars, get off scott-free? WPPSS should pay the debt on its outstanding bonds, not hand it off to the rest of us.

Not everyone was thinking like Congressman Weaver, however. City of Seattle officials were considering ways to help their fellow public utilities carry out a controlled termination of WNP 4 and 5, even though Seattle was not a participant on 4 and 5. In a controlled termination (provided there was enough money) WPPSS would acknowledge its obligation to contractors and suppliers, and everyone would try to work out some satisfactory form of payoff during a 12-month period. WPPSS felt that its costs in a controlled termination might run as high as $335 million—but in an uncontrolled termination, the costs might go even higher. In the latter case, WPPSS would tell its contractors and suppliers on WNP 4 and 5 that there was simply no money left for anything, would lock the gates to WNP 4 and 5, and would tell its contractors and suppliers to "get lost." Numerous lawsuits could then be expected against WPPSS, the 88 Participants, and other WPPSS assets such as WNP 1, 2, and 3, with the result that the Participants might end up owing even more money. Costs of an uncontrolled termination were estimated at $500 to $800 million, in addition to the costs of paying back all the bond debt. If the $300 million in funds were not available (from WPPSS or the 88 Participants) for a controlled termination during one-year period to pay bills when due, there would be a very serious risk of default plus the likelihood of an uncontrolled termination as well. In addition, lack of an orderly and well thought out termination could jeopardize the financing needed to continue construction of WNP 1, 2, and 3.

As Seattle officials and others worked to obtain support for a controlled termination of WNP 4 and 5, Oregon's Congressional delegation immediately went on record opposing any Federal "bailout" of WPPSS. U.S. Senators from Oregon, Mark Hatfield and Bob Packwood, said that the utilities would have to solve the WPPSS problem by themselves. As a result, in the days following WPPSS' termination, most of the 88 Participants in Washington, Oregon, and Idaho listened to Congressman Weaver and began looking for a "legal" way out of their indebtedness.

The first public utility group to publicly announce that it was looking for a way out of its contract with WPPSS was the Canby Utility Board, an independent governmental subdivision of the city of Canby, Oregon. Canby is a tiny, rural community about 40 miles south of Portland. The day before WPPSS officially terminated WNP 4 and 5, Canby Utility Board officials announced at a town meeting that because of WNP 4 and 5's termination, the Utility Board faced a debt of over $1 million a year for the next 35 years! The Board felt it had only one alternative. It was going to court to get out of its Participants' Agreement with WPPSS on WNP 4 and 5.

WPPSS Seeks Voter Approval for New Bond Sale

While some of the 88 Participants in WNP 4 and 5 were eyeing the possibility of filing lawsuits in attempts to get out of their Participants' Agreements, WPPSS was going about the business of completing WNP 1, 2, and 3. It didn't take WPPSS long to stub its

Making him a little nervous

Spokesman Review February 11, 1982; reprinted by permission.

toe again. In late January 1982, WPPSS angered state officials by announcing that it intended to seek voker approval in September to sell several hundred million dollars worth of new revenue bonds for WNP 1, 2, and 3. Voter approval would be required because of passage a few months earlier of Initiative 394.

Washington Secretary of State Ralph Munro was furious when he heard of WPPSS' plans, through the press. "I'm madder than blazes if this is true," he said. "It would require a separate voters pamphlet, which would cost $500,000 to $1 million [to print and distribute] and bust the budget of my office." (Initiative 394 required WPPSS to hire an independent firm to study the feasibility of the project before submitting the bond sale to voters and to publish the findings in a voter information pamphlet; holding the vote in September would mean the information could not be included in the pamphlet already planned for the November general election.)

Secretary Munro was also angered by WPPSS' choice of the NUS Corporation of Rockville, Maryland, to conduct the feasibility study. Once known as the Nuclear Utility Services Corporation, NUS Corporation currently had an $800,000 contract with WPPSS, and, in addition, had contributed $500 to the campaign against Initiative 394 the year before. People were saying that the WPPSS/NUS connection smacked of a conflict of interest. According to WPPSS' legal counsel, E. Craig Doupe, a September election was picked so that "we could take another crack in November" if disapproved in the September. In regard to selecting NUS Corporation to conduct the feasibility study, Doupe said that NUS' existing $800,000 contract was "minor" and that WPPSS' selection committee had not known about the $500 donation. Later, state officials, citing conflict of interest, disallowed use of NUS Corporation.

Seattle Searches for Funds for a Controlled Termination

While WPPSS was busily attending to WNP 1, 2, and 3, the city of Seattle continued its efforts to raise money for a controlled termination of WNP 4 and 5. Their effort was given a boost by the announcement of a $12.3 million loan from the Snohomish County PUD, one of WPPSS' staunchest supporters and owner of 13 percent of WNP 4 and 5. A few other utility districts were also considering making loans. WPPSS believed that it would need about $192 million during 1982 to pay contractors' claims and other bills associated with WNP 4 and 5 and that it would be able to contribute about $120 million of its own funds. WPPSS wanted the 88 Participants to raise $70 million through loans in 1982 and another $94.6 million through the life of the termination plan. It was hoping that the $70 million from the 88 Participants would be available by March so that the controlled termination plan could be implemented. Puget Sound Power & Light Company said that it would be unable to loan money for termination and, moreover, said that it would need the money it had loaned for mothballing returned.

Few other people were certain that the $70 million would be found. Four Oregon PUDs told WPPSS that the same law that prevented them from providing funds for mothballing also prevented them from making a loan for a controlled termination. A number of Washington Participants were also refusing to contribute to WPPSS' controlled termination fund, many citing the fact that no one, not even WPPSS, knew for sure what termination would really cost.

In spite of the many unknown costs for a controlled termination, 42 of the 88 Participants in WNP 4 and 5 finally agreed to make $70.5 million in loans to protect WPPSS against default and to help pay the first-year termination costs. The first installment on the loan payments, totaling $4 million, was due at WPPSS headquarters in Richland, Washington, on June 1, 1982. But by June 2, only 16 utilities had made payments, totaling $1.6 million. WPPSS staffers called many of the utilities and were told that the payments would be made. However, two small electrical co-operatives—the Fall River Electric Cooperative of Ashton, Idaho, and the Hood River Electric Cooperative of Odell, Oregon—said they would not be making payments, even though they had signed termination loan agreements. WPPSS officials said that they had no choice but to take legal action to force payment by the two utilities. "The amount they are withholding is relatively small," said Robert DeLorenzo, WPPSS' termination project director, "but they do have a contractual obligation." According to DeLorenzo, all late payments would be charged a 15 percent interest charge.

On June 19, 1982, WPPSS carried out its threat of legal action by filing lawsuits in Benton County Superior Court (eastern Washington State) against eight public utility districts: Hood River Electric Cooperative (Odell, Oregon); Fall River Rural Electric Cooperative, Inc. (Ashton, Idaho); Wasco Electric Cooperative (The Dalles, Oregon); Idaho Falls Electric Light Department (Idaho Falls, Idaho); Raft River Rural Electric Cooperative, Inc. (Rupert, Idaho); Unity Light & Power Company (Burley, Idaho); Rural Electric Company (Rupert, Idaho); and Lost River Electric Cooperative (Mackay, Idaho).

Although eight Participants had failed to make good on promised loans, WPPSS was able to raise $4 million in order to begin its controlled termination plan. All loans for termination would be repaid beginning in January 1983, when all 88 Participants in WNP 4 and 5 began paying off the $2.25 billion debt.

So Let 'em Sue

With the outcome of the controlled termination plan in limbo, the Participants were now wondering how they would ever be able to pay their share of WNP 4 and 5's $2.25 billion debt. Electric rates would have to rise immediately—and this prospect was causing the PUDs to have second thoughts about their obligation to repay the WNP 4 and 5 debt. The city of Centralia started looking at its Participants' Agreement with WPPSS and decided that it really never had borrowed money from anyone. WPPSS had borrowed money, but "all we did was sign an agreement to buy power," said Centralia PUD Commissioner Jack Gelder. "And that," the Centralia people said, "brings up another point: WPPSS signed a contract, but clearly didn't live up to its part of the bargain. It failed to build the plants on time or within budget." Centralia officials thought WPPSS should be sued, and they had talked to other cities about a lawsuit. Meanwhile, they were pretty much against any loans to WPPSS for a controlled termination. They knew that failure to pay Centralia's share of WPPSS' construction costs might result in WPPSS or someone else suing the city. Of that, the Centralia city commissioner said, "Let 'em."

You're Not Fighting Hippies Anymore

While the Participants debated whether or not to loan money to carry out the termination plan, irate ratepayers from all of the Participants were demanding that their commissioners, who got them into the WPPSS mess in the first place, resign.

At a Grays Harbor PUD commissioners meeting on February 2, 1982, about 100 ratepayers angry about their soaring electric bills confronted the commissioners and demanded that they resign. Many in the audience alleged that the PUD officials had sold them down the river through their zealous support of WPPSS' nuclear power projects. They complained about electric rates that had doubled in the past two years and (according to a study conducted by utility consultant R.W. Beck & Associates) were likely to double again in the next two years. "You're not fighting a bunch of hippies anymore," warned one ratepayer.

"And I don't intend to renege on our commitment," replied PUD Commissioner Jack Welch, referring to the PUD's obligation to help pay the costs of terminating WNP 4 and 5. "We intend to do everything we can to make sure that termination is controlled." The commissioners told the group that they were bound by contracts signed six years earlier to stand behind the defunct projects. The sentiment of the crowd was that the PUD should not pay another penny on WNP 4 and 5. "If anyone should take a beating, it should be the bond buyers," said one ratepayer. "You may not realize it, but the people in Grays Harbor PUD are a little wiser, a little stronger than to sit here and let you put us into debt," she added. Continuing, she told the PUD commissioners she was organizing a recall petition to oust the men from office. "You get the petition and I'll be the first to sign," Commissioner Arne Holm shot back.

The anger in Grays Harbor was typical of sentiment throughout Washington State. By February 1982, some 31 PUD officials sympathetic to WPPSS had been identified as recall candidates by an antinuclear group calling itself "Progress Under Democracy."

The Lawsuits Begin

In December 1981, the Springfield (Oregon) Utility Board (SUB) and several other Oregon utilities sought a Lane County Circuit Court injunction preventing them from participating in the WPPSS financing plan on WNP 4 and 5, claiming that its financial liability agreements with WPPSS violated Oregon State statutes as well as the Oregon Constitution and the Springfield City Charter. In addition, SUB's lawsuit charged that it had violated the law by pledging its utility revenues to meet its contract obligations instead of issuing revenue bonds. The lawsuit also alleged it had violated the city charter by not calling an election for approval of bonds to pay for Springfield's 1.8 percent share of the power output. Springfield's complaint also claimed that WPPSS had failed to live up to its agreement to build the terminated WNP 4 and 5, that WPPSS had not designed the plants properly, had failed to obtain "workmanlike" and timely performances from its third-party contractors, and had not administered its labor contracts properly and, by failing to audit financial records, had caused overpayment to contractors. These problems, the suit claimed, had caused enormous cost overruns at the plants, as well as delays in the construction schedule.

The potential gain from winning the suit would be that the SUB would be released from its estimated $57.8 million debt on WNP 4 and 5. While the Springfield Utility Board was suing WPPSS for breach of contract, the SUB itself was being sued in Lane County

Circuit Court by ratepayers seeking to block their utility from making payments to WPPSS. That suit claimed the SUB had violated state law and the city charter when it entered into a contract with WPPSS, so the contract should be declared void. The Springfield Utility Board believed that it had had the authority to enter such contracts, so it decided to fight the ratepayer suit.

They Were Our Daddy

Regardless of whether controlled termination of WNP 4 and 5 was successful, debt service payments on the $2.25 billion in bonds would have to start in January 1983. Plaintiffs in the Oregon complaints (the Springfield Utility Board and the other PUDs that joined in) were hoping to have a decision before that time as to whether municipal corporations like the SUB legally had to make those payments. Meanwhile, on April 26, 1982, 12 regional PUDs filed suit against WPPSS in Lewis County (Washington) Superior Court in an effort to escape paying millions in debt on WNP 4 and 5. The suit was dubbed the "Seduction Suit" by the "Dirty Dozen." Ten of the plaintiffs were located in Oregon, one was in Idaho, and one, the Oracs Power & Light Company, was in Washington. The suit asked a judge to determine if the 12 utilities, as well as the other 76 public utilities that had agreed to pay for WNP 4 and 5, could be held liable for the $2.25 billion debt since they had never received any power for their investment. In addition, the suit alleged that there was liability on the part of BPA for persuading (seducing) the 88 utilities to invest in the two nuclear plants in the first place. PUD officials said that they never would have agreed to built WNP 4 and 5 had it not been for BPA officials. "They [BPA officials] were our daddy," said Ray Rigby, attorney for the Fall River Rural Electric Co-op at Ashton, Idaho. "They said, 'Kids, you're not going to have power if you don't do this.' Dad says do it. We did it." Rigby and many others said one of the reasons for getting lawsuits in place was to show Congress, the Northwest Power Council, and state and Federal agencies that the WPPSS morass "is a national problem and won't go away on its own."

Edwin Schlender, manager of the Coos-Curry Electric Co-op in Coquille, Oregon, one of the so-called "Dirty Dozen," also said that the PUDs had been under heavy pressure ("coercion," said the Fall River Co-op) from the Rural Electrification Administration (REA), banker for all the electric co-ops, and the Co-operative

Making the final concrete pour at the top of WNP 1 containment building shortly before the project was mothballed, circa May 1982.

Finance Corporation, a bank owned by co-ops around the country, to shut up and pay the WNP 4 and 5 debt. (Later, in a report by the U.S. General Accounting Office dated July 30, 1982, REA would be criticized for "directing its borrowers to pay their debts to the WPPSS and implying their financial standing with REA could be affected if their obligations on WNP 4 and 5 are not met.")

While the lawsuits against WPPSS continued, WPPSS was being urged by 18 Washington state legislators to mothball WNP 3, the remaining power plant at Satsop. A Washington State University report had indicated, and BPA forecasters had confirmed, that the power from the nuclear plant might not be needed as soon as had been predicted. "An orderly mothballing of

WNP 3 would be preferable to the management-by-crisis which has accompanied action on WNP 4 and 5," said the legislators. WNP 3 was about 47 percent complete and was scheduled to come on line in 1987. About $1.13 billion had already been spent. WPPSS spokesman Steve Irish opposed mothballing WNP 3: "Stopping and resuming construction, together with regulatory changes during a construction delay and inflation, would all add costs to the plant which is expected to cost $4.5 billion when completed."

Uphold Your Contractual Obligations

In May 1982, WPPSS began reminding the 88 Participants in WNP 4 and 5 of their financial obligation to repay the $2.25 billion debt. The Grays Harbor PUD, which would have to pay back about $300 million over a 35-year period, sought a legal opinion on the indebtedness issue from Washington State Attorney General Ken Elkenberry. Their question centered around a state law requiring a PUD to get voter approval on PUD indebtedness exceeding three-quarters of one percent of the value of taxable property in the county. The people in Grays Harbor wanted to know whether that law applied to them. Elkenberry thought it didn't. He said that the state law applied only to debt incurred by the PUD through its own bond issues. In the case of WNP 4 and 5, the PUDs hadn't issued the bonds—WPPSS had. Therefore, said the Attorney General, public utility districts had not needed to secure voter approval before signing the Participants' Agreements with WPPSS.

Chemical Bank's Complaint

On May 19, 1982, the bond trustee for WNP 4 and 5, Chemical Bank of New York, filed a complaint in King County (Seattle) Superior Court asking the court to rule that the Washington State Participants in WNP 4 and 5 were obligated to pay off $7 billion in principal and interest over the next 30 years even though the two projects had not been completed. In effect, Chemical Bank, acting in behalf of bondholders, was seeking an early court decision on the legality of the Participants' Agreements, specifically on their "hell or high water" clauses.

Let's Abolish WPPSS

In July 1982, Steve Zemke of the Don't Bankrupt Washington Committee, began hinting that the Committee was working on a new initiative that would have the effect of abolishing WPPSS. Zemke's group had succeeded in getting voter approval of Initiative 394, which required a public vote on bond sales for financing major energy-generation projects. At the time he was campaigning for Initiative 394 in mid 1981, Zemke maintained that the initiative wasn't designed to dismantle WPPSS. Now his group was considering an initiative that would do just that, because of action the Washington State Legislature had taken that resulted in making WPPSS' Executive Board less responsive to voters and ratepayers. According to Zemke, WPPSS' 11-member Executive Board now had six members appointed from outside WPPSS member utilities. And of the five members representing Pacific Northwest public utilities, only three were elected public utility commissioners. Said Zemke: "The whole thing of public power where people are elected to serve has now been made a sham. There is an extreme sense of frustration among ratepayers of the state because they feel they have no control over WPPSS." Due to this lack of control, Zemke was considering an initiative to repeal the state law allowing the formation of joint operating agencies—the measure under which WPPSS was established. "We would like to shut it down [WPPSS] and declare the property surplus and then sell it to the highest bidder," said Zemke.

Legal Complaints Draw Wall Street Warning

By mid-September 1982, Wall Street analysts were looking at the legal challenges related to WNP 4 and 5 and becoming worried. Eileen Titmuss Austen, vice president and manager of municipal research for Drexel Burnham Lambert, Inc., wrote in the July edition of *Bond Buyer* magazine that "a default on bonds for terminating WNP 4 and 5 is not out of the question." She cited WPPSS' difficulty in collecting $70 million in loans for termination and noted that that problem was small compared with the potential problem of the utilities not making interest payments scheduled to begin in January 1983. Austen noted that WPPSS was resorting to legal action to collect promised loans intended to carry out a controlled termination of WNP 4 and 5. "If enough participants fail to make payments sufficient to cover the next six month period," Austen said, "a default could follow. Given the combined effects of increased financial strain to some participants and the legal questioning of many others, we believe that such an occurrence is not out of the question."

Oregon Utilities Freed of WPPSS Debt and Investors' Worst Fears Realized

On October 7, 1982 Judge George Woodrich of the Lane County (Oregon) Circuit Court absolved the Springfield Utility Board's ratepayers of their $124 million share of the WNP 4 and 5 debt. Judge Woodrich said the Springfield Utility Board's contract and those of 16 other Oregon utilities were invalid because the utilities lacked authority to enter into them. The contracts were contrary to Oregon law, and some city charters as well, said Woodrich, because they involved the production of energy outside utility-district boundaries and failed to give the Oregon utilities actual ownership in the electric generating plants. The Springfield Utility Board, and the other utilities, did not have an ownership interest in WNP 4 and 5, Woodrich noted, but merely had contracted for future power purchases. Furthermore, Woodrich wrote, the contracts were void because the utilities had exceeded their charter debt limitations and because they had failed to get approval from ratepayers before signing the Participants' Agreements.

Investors and others in the municipal bond community were not prepared for Judge Woodrich's decision. A group of people (utility commissioners) had signed agreements that supposedly provided financial backing for a power project. On the strength of those agreements – which stated that the bonds would be paid off even if things fell apart – the bonds were sold to another large group of people and organizations. It now appeared that at least some of the signers – a group of Oregon utilities – were nothing more than a group of unauthorized borrowers. Now many people throughout the United States who had loaned money on the strength of those agreements were left holding the bag.

Oregon Congressman James Weaver called Judge Woodrich's decision in the Springfield Utility Board case "one of the greatest victories I've ever had." According to Weaver: "The specific victory to me is I said the contracts were no good and would not stand up – that's been my position on that – and I was vindicated."

Earlier Weaver had claimed credit for alerting Wall Street to WPPSS' contractual and financial problems. Weaver said he had warned utility officials and New York securities analysts to "sell those bonds at your own risk" and that "Oregon's not going to pay them because we're going to court and if the court finds those contracts invalid, we don't owe the money." In fact, the two men who had filed the suit against WPPSS and the Springfield Utility Board, Peter DeFazio and David Dix, were on Weaver's Congressional Staff. DeFazio was Weaver's director of constituent services in Eugene, Oregon, and a candidate for commissioner in Lane County, where the Springfield Utility Board is located. DeFazio and Dix said that they had filed the complaint in December 1981 out of personal interests, and not because of their connection with Congressman Weaver. According to DeFazio, the Springfield Utility Board would not admit that it had made a mistake and would not take any legal action on behalf of its ratepayers. DeFazio said that he had tried numerous times in November and December 1981 to persuade the SUB to file suit against WPPSS. But the Board rebuffed his suggestion after its legal counsel advised that there was nothing wrong with participation in WNP 4 and 5. On December 23, 1981, DeFazio, along with 25 Springfield ratepayers as co-plaintiffs, filed suit against their utility and WPPSS. DeFazio said that he filed "out of complete frustration with the Utility Board.

As for his role in the lawsuit, Congressmen Weaver said: "I encouraged them [DeFazio and Dix] to do it. I gave them permission to do it as members of my staff. If I had been a SUB customer, I would have joined that lawsuit. I gave them the impetus. It was their decision, but they knew I wanted to sue." Weaver doubted that Oregon's credit would be harmed by the court's ruling. "If we owed the money and didn't pay, that would hurt us badly," said Weaver. "But if we didn't owe it, if those contracts are invalid, and we got swindled into this—which is in effect what we did—then Springfield just got itself absolved of a $125 million debt, through our lawsuit. They don't owe it anymore."

Weaver said the lawsuits would have been avoided if the 88 Participants and/or other Congressmen had listened to his plans to require a vote of the people before incurring the WPPSS debt. "[WPPSS and the 88 Participants] wrote the phoniest contracts imaginable because they didn't want to go to the people," Weaver continued saying that in recent court decisions a bond attorney admitted that he knew the contracts were against Oregon law. "Very frankly, what they've been doing in this whole mess for the last 15 years is against the law," Weaver said. "Now they're hoisted on their own petard."

Members of the Oregon "Irate Ratepayers" organization supporting the lawsuit were jubilant, too. "I think it's fantastic, don't you?" asked a member of the Irate Ratepayers of Clark County. The woman said her group contended that the Clark County PUD had violated the Oregon State constitution by exceeding a debt limitation provision without a vote of the people. "This is a confirmation of everything the Irates have fought for from Day 1," she said.

Stunned by Judge Woodrich's favorable ruling on behalf of the Springfield ratepayers, WPPSS' Executive Board announced that it was giving Managing Director Robert Ferguson 30 days to find a way of paying off the $2.25 billion in bonds issued to finance WNP 4 and 5. It also voted, the Executive Board, on the recommendation of the Participants' Review Committee, to pay $60,000 to Shearson/American Express, Inc., an investment firm, to conduct a 30-day feasibility study of a previously proposed refinancing plan for WNP 4 and 5. Under the rather complicated plan Shearson was to study how the Federal government could subsidize the debt by, in effect, granting WPPSS a $1.5 billion low-interest loan via BPA. WPPSS would place the proceeds of the BPA loan in a special escrow account, which would be invested commercially at higher interest rates. The escrow would become collateral for the outstanding bonds for WNP 4 and 5, giving them more security, and the extra interest earnings would be used to help pay off the bonds as they came due over the next 30 years.

The plan would have left WPPSS in debt to BPA but would cut the overall obligation for WNP 4 and 5 by half. Congressional approval and, certainly, BPA cooperation would have been necessary. It was anticipated that serious political obstacles would have to be overcome in order to obtain the WPPSS/BPA loan, but WPPSS Board members felt it was their last hope of avoiding default and therefore was certainly worth trying. They remained a bit optimistic for a long-term solution—and they got it in the form of a promising pre-trial ruling on Chemical Bank of New York's suit.

"Utility Districts Must Pay Debt"—Judge Coleman

On October 15, 1982, King County Superior Court Judge H. Joseph Coleman, in a pre-trial ruling on Chemical Bank of New

York's complaint, stated that the Washington State Participants were obligated to pay their shares of the debt on WNP 4 and 5 unless the court, in further proceedings, determined that the contracts with WPPSS were invalid because of misrepresentation during their formation. The 23 WPPSS (Washington) members were stunned. Throughout the pre-trial hearings, the Participants had argued that their Participants' Agreements with WPPSS were unfair because they required payment even if nothing was delivered, and they argued that as in Oregon, ratepayers should have been allowed to vote on the contracts. Chemical Bank, as the bondholders' trustee, argued that the Washington Participants had entered into the agreements in good faith, with their eyes open. They hadn't had to sign the agreements, but they did, and now they should honor them.

Judge Coleman noted that the legal arguments in the case were "detailed and sophisticated—a prodigious effort." "But," he continued, "a thorough reading of the contracts permitted only one conclusion, and that was that the participants undertook to pay the debt service and all other costs for [WNP 4 and 5] whether completed or not."

Coleman dismissed several other arguments raised by the utilities, including one that they lacked authority to enter into agreements about the nuclear plants. With the possible exception of a single irrigation district, "the participants had legal authority to enter into the agreements," he ruled.

Although the pre-trial ruling favored Chemical Bank, the case wasn't over. Coleman had not ruled on the validity of the contracts themselves, but only that the utilities must pay if the contracts were found valid. Moreover, Judge Coleman's decision applied only to Washington State Participants. Coleman said he would rule on the validity of the contracts themselves at a trial scheduled to begin on November 29. (At the request of attorneys for the utilities, Coleman later delayed the trial to January 10, 1983, in order to give the utilities and their attorneys enough time to prepare.) In the meantime, Coleman ordered the participating utilities to place their share of the debt in an escrow account until such time as the overriding obligation question was decided. He felt it would not be fair to the Participants if they sent their ratepayers' money to WPPSS, it was spent by WPPSS, and then the courts found the utilities didn't have to pay WPPSS after all. Immediately after Coleman's decision, the Washington Participants had appealed to the Washington State Supreme Court. The high court said it would review the appeal, and Judge Coleman's trial on the validity of the contracts was indefinitely postponed. The wait might be long, for the high court would have to wade through some 7,000 pages of documents.

No Way of Forcing WPPSS into Bankruptcy

Regardless of the final outcome of the Chemical Bank lawsuit, WPPSS would have started paying interest on WNP 4 and 5's debt by January 25, 1983, if it were to avoid default. If default occurred, said WPPSS officials, they could be headed for bankruptcy and receivership, and WNP 1, 2, and 3 might be jeopardized as well. WPPSS could not be forced into involuntary bankruptcy, WPPSS officials pointed out, because under Washington law a public agency like WPPSS can be in bankruptcy only by putting itself in bankruptcy. Some critics of WPPSS felt that bankruptcy might not be such a bad idea. U.S. Senator James McClure (R-Idaho) told a news conference in Spokane that bankruptcy might be the best way for WPPSS to deal with its financial crisis. WPPSS Managing Director Robert Ferguson, while declining to be specific, hinted that WPPSS was considering bankruptcy, among other options. At the same time, he insisted, the main priority was for WPPSS to live up to all of its obligations.

Snohomish District and Others Sue Bonneville for Coercion

The Participants were considering their options, too. With the Oregon Participants now off the hook in having to pay WPPSS back on WNP 4 and 5's debt, three Washington PUDs—Snohomish, Clallam, and Clark Counties—decided to file a lawsuit to rid themselves of their share in the debt, too. Their suit was against Bonneville Power Administration, and at its heart was the need for power. (The suit was filed on January 18, 1983, in the U.S. Claims Court, Washington, D.C., which handles all claims against the Federal government.) According to Ed Hanson, attorney for the plaintiffs, "the need for WNP 4 and 5 was created by BPA itself after [it] signed new power contracts with regional aluminum industries." Hanson claimed that BPA forced local utilities to build WNP 4 and 5 while it was selling industry power it didn't have. At the same time BPA was signing those industry contracts, it was also participating in studies that indicated BPA was running out of power. These accusations angered the regional aluminum industry, which claimed that the charge was totally inaccurate. "The theory that somehow the direct service industries are responsible for WNP 4 and 5 is just absurd," said Brett Wilcox, spokesman for the aluminum industry. The legal suit sought a money judgment against BPA and the Federal government and, if successful, would lift the debt on WNP 4 and 5 from the backs of all 88 Participants and shift it to Federal taxpayers.

WPPSS Bills Participants for First Payment

Under agreements with WPPSS, the 88 Participants were to begin making payments on the $2.25 billion debt on January 25, 1983. WPPSS mailed out the first bills in late October 1982, asking those who did not intend to pay to so inform WPPSS by November 3. The total amount requested was $162 million, enough to cover the first half of 1983. Some of that money was needed for interest payments, some would be used to pay off contractors and suppliers, and some would go to repay the earlier loans from utilities for WPPSS' controlled termination of WNP 4 and 5. The largest single bill, for $21 million, went to the Snohomish County PUD, which owned the largest single share of WNP 4 and 5, 13.051 percent—and which was about to file suit against BPA and the Federal government to rid itself of the debt.

By early November, only five Participants—the four Oregon municipal utilities of Drain, Bandon, Cascade Locks, and Canby and the Northern Wasco PUD—had informed WPPSS that they had no intention of paying their shares of the bill for WNP 4 and 5. (Only several weeks earlier, Judge Woodrich, presiding over the Lane County Superior Court in Eugene, Oregon, had reaffirmed his earlier Lewis County Circuit Court ruling that 11 Oregon utilities were not required to pay their contracted debt on WNP 4 and 5 because they had no authority to become involved in the two projects in the first place.) Since the Participants were not legally required to inform WPPSS of their intentions, by notifying WPPSS these utilities were extending a "courtesy." WPPSS did not know what the other Participants would do—and that was a crucial question, because these five Oregon utilities owed only about 3 percent of the total debt on WNP 4 and 5.

While WPPSS was wondering what the other Participants

would do, the Shearson/American Express study of a plan to refinance the debt on WNP 4 and 5 died quietly, on November 12, when WPPSS failed to renew Shearson's contract. The Participants' Review Committee, which backed the study, felt the refinancing plan was a long-shot gamble, if not impossible. Joe Custer, a Participants' Review Committee member, said: "To put a program like this together you need a strong consensus in the region for support, and there isn't that consensus at this time." Indeed, many of the 88 Participants in WNP 4 and 5 had gone to court to have their own contracts declared invalid, and, few of the 88 were interested in a refinancing plan. The Participants' Review Committee did not scrap the refinancing plan entirely, but merely placed it on hold. "The utilities may reach a time when they will again look for ways of reducing the $2.25 billion debt through a refinancing plan," said Custer. "So the Participants' Committee may again turn to Shearson for financial advice." WPPSS officials, friends, and supporters were confident while they waited for King County Superior Court Judge H. Joseph Coleman's final ruling on the validity of the Participants' Agreements. At the time, WPPSS was still hoping to recover from its financial problems on WNP 4 and 5 by its own efforts and the efforts of its 88 Participants. From time to time, newspaper reporters would ask U.S. Senator Henry Jackson (D–Washington) if WPPSS had requested any Federal aid. In late October he told reporters that "there had been no formal requests from WPPSS to the Federal government for help." He believed that the state and WPPSS should examine markets for electric power in the Pacific Southwest.

By late November 1982, few of the 88 Participants in WNP 4 and 5 had communicated with WPPSS about their intentions—and WPPSS officials were getting nervous. They decided in early December to force the silent Participants to make some sort of move toward repayment by taking the matter to court. They hoped to obtain an order forcing the Participants to raise their electrical rates so they could make payments, and then to report, by December 23, on how they planned to make them. Attorneys representing the utilities objected to the proposed order, so WPPSS withdrew its request and instead sought an order requiring only information on how the payments would be made. King County Superior Court Judge H. Joseph Coleman declined to issue such an order, saying that the issues raised might be settled later when he ruled on whether the contracts between WPPSS and their participants were valid.

Following WPPSS' unsuccessful effort to force the Participants to make some sort of move toward repayment, Washington

Governor John Spellman began a series of closed-door meetings with aides and business leaders to examine what the state would do if WPPSS could not pay its bills on WNP 4 and 5. Among the participants were George Weyerhaeuser, chairman of Weyerhaeuser Company, Bud Coffey, a Seattle-based lobbyist for Boeing Aircraft, Robert Graham, former Seattle Chamber of Commerce president, Larry Kenney, secretary-treasurer of the Washington State Labor Council, and Bob Dilger, executive secretary of the Washington Building & Construction Trades Council. Also at the meeting were Seattle attorney William Appel, Northwest Power Planning Council official Chuck Collins, and staff officials/members from the Governor's and attorney general's offices. The outcome of these sessions in early December 1982 was a campaign aimed at persuading WNP 4 and 5 Participants to pay their debts. Also discussed was a tax on all the utilities in the state. In an interview with the Seattle *Post-Intelligencer*, the governor was quoted as saying: "I don't think there's a way any ratepayer or district in the state is going to get off the hook." His comments immediately raised public outcry and criticism. An outspoken electric-ratepayer group, the Tacoma Light Brigade, said it would fight any statewide utility tax: Aides to the governor later said that he was misunderstood, that he had not meant to imply that the $7 billion debt left by the termination of WNP 4 and 5 should be borne by all the state's electrical utilities, but only by the 88 utilities owning shares of the plants. Kate Heimbach, the governor's deputy press secretary, said there was "a problem of semantics; the governor was referring to only the 88 participant ratepayers and not all ratepayers."

Former Washington Governor Dan Evans, who had signed the original documents allowing WPPSS to carry out its nuclear program, held out little hope that the utilities would be able to escape the burden of the WPPSS contracts. The former three-term governor was speaking from his position as chairman of the Pacific Northwest Electric Power and Conservation Council. He thought that one of the many ideas being discussed, regionalization, and specifically selling WPPSS power in California to spread the burden and make the huge debt more bearable, was a dream that would never come true. "It's a long shot," he said, "because only BPA could be the buyer of the power. And BPA would have to assert it's really needed, that WPPSS power is price-competitive, and that alternative sources aren't available." Evans noted that demand for energy was decreasing. "It isn't possible to compete with California-generated power," he pointed out. And just conserving power was proving to be a new "source," as the people of Seattle had learned. The upshot of the WPPSS/utilities dilemma, Evans predicted, would be that the contracts would be upheld and that the 88 utilities would be made to follow through on their contract obligations.

Evans discounted the importance of Oregon Judge Woodrich's decision that the Oregon utilities had no authority to enter into contracts with WPPSS. If the Washington utilities' contracts were upheld, as he felt they would be, then, he admitted, WPPSS' Washington Participants would be in varying degrees of economic distress—and perhaps some of the smaller ones would not be able to pay. "They'll be helped," Evans assured them, "but at this time I don't know how that help will be provided. I don't know how it'll be done."

The former governor mused about the origins of the WPPSS mess. "After all, I signed those original documents as govenor," he smiled.

> But, initially when the WPPSS people came to me with the proposition that we needed all this additional power, they were using a six percent demand growth rate. I thought that was too high. I sent 'em back to refigure. They came back with a lesser figure which indicated the plants would still be needed. Finally, I asked 'em to base their figures on a three percent rate. They did. The plants were still shown as needed.

In early December, utility lobbyists and state legislators in Washington State also were discussing measures to ease the threat of default by the 88 Participants and WPPSS. Most felt that there was little chance Washington's legislature would act unless Idaho, Oregon, and Montana took similar action. That left only one good idea—a surcharge on all electric utility bills. Other ideas that had been reviewed by legislators included an "export tax" on energy generated in the State of Washington and sold to other states in the Pacific Northwest. In the end, state legislators felt that none of their ideas would pass the legislature. They felt they were pretty much powerless without the cooperation of the other three states.

Also in early December 1982, WPPSS Managing Director Robert Ferguson announced that he was resigning, effective June 1983. Citing continuing health problems, Ferguson said he had not recovered fully from open-heart surgery the preceding spring. Ferguson was known as a "first class construction ramrod" when he was hired as managing director in 1980. But he now realized that

WPPSS needed more than a construction whiz. WPPSS had, he thought, pretty much overreached itself in trying to build five nuclear power plants simultaneously.

Six weeks later, he would be supported in his decision by the three-member panel formed by the governors of Oregon and Washington to examine the impact of WNP 4 and 5 on the economy, the utilities, and the citizens of the Northwest. They recommended that the plants be mothballed – and that the mothballing be done quickly, to avoid a total "financial meltdown."

When Ferguson, in January 1982, had recommended WNP 4 and 5 be terminated, the 88 Participants had agreed. But they also blamed Ferguson and his staff for prematurely calling for a construction moratorium. He, in turn, blamed the utilities for failing to act quickly to save the plants. At first he had called for a construction "moratorium" in May 1981 on two of the plants making the 88 Participants furious. (In fact Ferguson had made the decision without consulting the Participants and when he eventually did talk with them, it was only 24 hours before he went public with his moratorium plan.) Ferguson had pleaded with the Participants for an orderly construction slowdown, for mothballing the plants for up to five years. When they were unable to agree on a plan, Ferguson had become increasingly frustrated. Time and again, he had warned the utilities that the money was running out and there might be no choice but to terminate the projects for good unless they acted quickly.

After January 1982, things were seldom the same between the managing director, the Board of Directors, and the 88 Participants. In the end, the stress of the WPPSS rescue effort proved to be too much for one man, even for Ferguson. Most of the people associated with WPPSS and the Participants agreed that it might be better to have someone new come in and guide WPPSS through its increasingly troubled times. (Stanton "Nick" Cain, chairman of WPPSS' Executive Board, said he had an idea of what kind of person WPPSS needed: "We absolutely have to have a professional manager, someone who knows how to deal with all the problems ... and it wouldn't hurt if he were a miracle worker.")

Although he was now a "lame duck," Ferguson continued working behind the scenes to keep WPPSS from going broke. He was convinced that a long-term plan to sell Washington's excess power outside the region would allow WPPSS to "manage its way out of [technical] default." Although Dan Evans and other regional leaders dismissed "regionalization" as unworkable, Ferguson continued to explore its possibilities, believing that the demand in California, for example, might be great enough to absorb the power produced by WPPSS. To this end, he asked Charles F. Luce, a Portlant, Oregon attorney and retired chairman of New York City's Consolidated Edison, to contact California electric utilities. Luce contacted the Sacramento Municipal Utility District, Pacific Gas & Electric Company, and Southern California Edison Company, which, Ferguson believed, might be interested in purchasing WPPSS-generated power under 5-, 10-, or 15-year contracts. Luce told the three utilities he contacted that if enough California utilities agreed to power contracts, then WPPSS could obtain necessary financing and complete the projects. Once the plants were finished, said Luce, WPPSS' electricity could be blended with power from cheaper sources (BPA) in the Pacific Northwest, then sold to California utilities. Several California utilities were interested, but they feared that the power would be very expensive. They were not ready to make any commitments, at least not until the price of the power was established.

Governor Spellman, happy over the news that California utilities might be interested in WPPSS electric power, said the deal would hinge on whether the Federal government could agree to purchase the two abandoned plants. He still maintained, however, that the 88 Participants, which in effect owned the two terminated plants, must begin paying their $2.25 billion debt. If Congress later agreed to a bail-out plan, said Spellman, then the 88 utilities' debts would shrink. This was the first time that the governor admitted he was trying to convince Congress and the Reagan administration to buy one or both of the nuclear plants.

Ferguson said that regardless of the eventual outcome of Luce's sales pitch in California, his remaining time with WPPSS would be largely devoted to finding ways of paying WPPSS' bills and locating a buyer (or buyers) for WNP 4 and 5. He felt that the next six months would determine whether WPPSS went broke or moved ahead to finish at least three of its nuclear plants. (In March 1983, the California Public Utility Commission said that it did not want utilities in that state to underwrite completion of Washington's two terminated nuclear power plants. "We've got our own nuclear problems down here and we don't want to import any more," said Leonard Grimes, Jr., president of the California PUC.)

In mid-January 1983, Governor John Spellman announced that he had hired Lester B. Knight & Associates, Inc., of Chicago, an international management consulting firm, and Bonnwell &

Company, specialists in tax-exempt securities, to study the financial impact default on bonds sold by WPPSS for WNP 4 and 5. The purpose of the 60-day "independent" study, which would cost $95,000 and was to be financed out of the governor's emergency fund, was "to resolve conflicting fact and interpretations surrounding the possible impacts of default on WPPSS, participating and nonparticipating utilities, on local units of government, and the economy of the Pacific Northwest." Presumably, this study would help clear the air. However, the governor's choice of study consultants added even more woes to his beleaguered refinancing efforts. On January 21 a top official of Lester B. Knight & Associates acknowledged that the firm was a controlling stockholder in two companies that owned $350,000 worth of WNP 4 and 5 bonds (KEAG and LBK Investment Company). Governor Spellman's press secretary, Paul O'Conner said the $350,000 holding was "peanuts" compared with the outstanding debt and hence was irrelevant. Initially Spellman had said that he had selected Lester B. Knight & Associates because he didn't want anyone who was dealing with WPPSS bonds or doing business with WPPSS. He also had noted how he had gone out of the Pacific Northwest and the New York City financial community to find a firm with no WPPSS connections. The consulting company issued a statement saying that "these bonds were sold or transferred to third parties before the execution of the contract." Spellman remained unconvinced that Knight's holdings would undermine the credibility of the upcoming study.

The WPPSS Participants Ignore Debt Payment

As WPPSS' January 25, 1983, bond payment deadline approached, there had been some indication that several of their 88 Participants wouldn't pay their shares of the more than $162 million due. For example, on December 15, the Rural Electrification Administration (REA) reversed its earlier tough stand and granted 17 of its co-op member Participants in WNP 4 and 5 relief from making any immediate payments to WPPSS until the complaint by Chemical Bank against the 88 Participants had been settled. Idaho Congressman George Hanson had bitterly complained to REA that its earlier stance "is unconscionable in the face of massive litigation now proceeding in several cases contesting the validity of the contract under which these co-ops are alleged to be responsible for the WPPSS bonds, principal and interest." REA had threatened to withhold funds from all Northwest energy cooperatives that refused to pay the bills issued by WPPSS on WNP 4 and 5.

Come payment day, only 2 of the 88 utilities, the Douglas County PUD (Washington) and Wells (Nevada) Rural Corporation, made payments to WPPSS, and those two payments totaled less than $10,000. All the other Participants in WNP 4 and 5 ignored the deadline, though a few did place the money owed in special reserve funds under their own control, pending the outcome of the Chemical Bank lawsuit. One of these was the Snohomish County (Washington) PUD, which owed the biggest share of the two terminated plants and which on January 18, 1983, filed suit against BPA seeking to rid itself of the debt. "It was the first time in the Snohomish County, PUD's history that the commissioners said no to WPPSS," said Gordon Rosler, leader of Fair Use of Snohomish Energy, a local ratepayer group. One of the Snohomish County PUD's partners in the lawsuit against BPA, Clark County PUD, which owned 9.6 percent in WNP 4 and 5, did not pay—and didn't even bother to collect the money due WPPSS.

The refusals to pay left WPPSS far from the sum needed to make bond payments and settle contracts on WNP 4 and 5's termination. It could contribute about $47 million left from previous WNP 4 and 5 bond sales, but that would allow payments to be made only until mid March. Maybe it was irrelevant anyway: Unless the 88 Participants came in with at least $94 million by July 1, WPPSS would have to face default.

"Lawyers Relief Act of 1982"

Things were not looking good for WPPSS. Only two of the 88 Participants had met their debt obligation. The Oregon Superior County Courts had let the Oregon Participants off the hook on WNP 4 and 5. The Idaho Supreme Court had ruled that Idaho PUDs that owned shares of WNP 4 and 5 could not raise their electric rates in order to pay interest and principal bills WPPSS had sent them. The Rural Electrification Administration had backed off from forcing the co-ops to pay WPPSS. As long as King County Superior Court Judge Coleman delayed his ruling on the validity of the contracts between WPPSS and the Washington Participants, there remained a doubt about when, if ever, the Participants would begin paying interest and principal on the two terminated plants. The longer all this dragged on, without the matter being settled, the closer WPPSS was coming to default on the $2.25 billion worth of bonds.

As the WPPSS affair continued on into 1983, more and more lawyers began working on WPPSS-related cases and were believed to be making millions of dollars. The amount of money to be earned in legal fees prompted people to begin calling the WPPSS legal tangle the "Lawyers Relief Act of 1982." The most eagerly awaited of all the trials continued to be Judge Coleman's final ruling on the validity of the Participants' Agreement. (Complicating things, the utilities had appealed Coleman's pretrial ruling to the State Supreme Court, and Coleman's trial in Superior Court could not continue until the Supreme Court had acted on that appeal.) In addition, numerous lawsuits were being filed by bondholders moving to protect their investments in WNP 4 and 5. In one such suit the bondholder, who had purchased $50,000 in bonds in April 1981, asked the Seattle Federal court to declare WPPSS in default, to order the 88 utilities that had sponsored WNP 4 and 5 to pay the debt, and to award undetermined damages to the investors. Defendants in the complaint were WPPSS, the 88 utilities, R.W. Beck & Associates (a Seattle engineering consultant retained by WPPSS), United Engineers & Constructors, Inc. (architect-engineer on WNP 4), Ebasco Services, Inc. (architect-engineer on WNP 5), and the bond underwriting firms of Salomon Brothers, Merrill Lynch, and White Weld Capital Markets Group. The complaint charged that WPPSS, Beck, United Engineers, and Ebasco had "made untrue statements" about the cost and completion schedules on WNP 4 and 5 when the bonds were sold and had "omitted to state material facts." The complaint further charged that in "acting with intent to defraud or in reckless disregard of the truth" the defendants had included "totally inaccurate cost projections, use projections and completion dates" for the nuclear plants in official statements published for bond sales. As a result, the plaintiff said, he, had purchased bonds "at artificially inflated prices." He said that the value of the bonds had dropped substantially because the projects had been terminated. The suit was filed as a class action in behalf of other purchasers of WNP 4 and 5 bonds. In response to the suits, Harold Hill, head of the Washington PUD Association, said: "It's the sort of action which emphasizes that now more than ever we in the region must pull together if we are to find a fair and equitable solution." Hill felt that without some sort of out-of-court settlement, the region would face a decade of suits and countersuits.

A Legal Tangle Only a Lawyer Could Love

On June 15, 1983 the Washington State Supreme Court handed down what has become a very controversial decision (Chemical Bank v. Washington Public Power Supply System). All the Washington participating utilities were relieved of their WPPSS debt. The decision left bondholders of WNP 4 and 5 with a huge potential loss. However, the legal war was just beginning. Bondholders had not given up the search for someone to pay off the debt. As of July 11, 1983, at least 13 bondholder lawsuits had been filed, and more are certainly expected. "Everyone who has ever been near one of the parties is looking at a suit," said Michigan bond lawyer Judson Werbelow. In addition to WPPSS, its members, and the 88 utilities that promised to pay for the two terminated plants, the list of those being sued includes underwriters and an array of professionals who worked on issuance of the bonds. A "who's who" of defendants currently caught in WPPSS lawsuits includes:

Bond Underwriters

Merrill Lynch Pierce Fenner & Smith
Prudential-Bache Securities
Salomon Brothers
Smith Barney, Harris & Upham

Rating Agencies

Moody's Investors Service
Standard & Poor's

Law Firms

Houghton, Cluck, Coughlin & Riley (Seattle)
Wood & Dawson (New York City)

Engineers and Contractors

R.W. Beck & Associates
Ebasco Services (a division of Enserch)
United Engineers & Constructors (a division of Raytheon)

Financial Advisor

Blyth, Eastman, Paine & Webber

The legal actions filed thus far are fraud cases brought under securities laws. They claim that when the bonds were issued, buyers were misled about such critical things as estimated cost and the utilities' legal authority to sign WPPSS contracts. Notes one expert: "It should have been known that it would take a lot more money to build WNP 4 and 5 than it was represented it was going to take."

The heart of the WPPSS bondholders' securities fraud cases is the charge of misrepresentation. One of the suits, for example, claims that Merrill Lynch Pierce Fenner & Smith and Salomon Brothers knew about or "recklessly" disregarded information concerning "the serious and deteriorating problems in the management, construction, and financing of the projects and acted to conceal the same." Another case asserts that WPPSS' New York bond counsel, Wood & Dawson, failed to tell purchasers that it had approved the legal authority of only 72 of the 88 participating utilities "because of possible invalidities" concerning the remaining 16. However, the underwriters say there was no fraud. They argue that securities laws permit them to rely on the opinions of other professionals, such as engineers and lawyers. But more more difficult, say the experts, is that WPPSS' bond counsel lawyers now face the difficult task of having to argue that they correctly analyzed the utilities' authority despite the fact that Washington's highest court has overruled them. Said Wood & Dawson partner Thomas L. Poscharsky: "We think their decision is a wrong one."

13

Chemical Bank vs. Washington Nuclear Plants 4 and 5 Participants

By April 1983, the Participants in Washington Nuclear Plants 4 and 5 had brought the Washington Public Power Supply System to the brink of default. WPPSS and Chemical Bank of New York were waiting for a clear statement from the Washington State Supreme Court that WNP 4 and 5 Participants in Washington must pay their bills. Most of the Participants were telling WPPSS that they wouldn't even consider paying until the Washington State Supreme Court told them they had to.

Time was running out for WPPSS. It had to come up with more than $50 million by the middle of May to pay the interest on bonds and other costs. If it was forced to admit that it couldn't meet the payment, default proceedings could be implemented at its next Executive Board meeting, on May 13. Because it appeared that the Washington State Supreme Court might take a lot of time, time they didn't have, WPPSS officials appealed to the high court to allow Judge Coleman to proceed with his planned trial to determine who should pay the massive debt. The high court gave a terse response to WPPSS' request: Until the review was completed, Judge Coleman could not proceed with the trial.

The Washington State Supreme Court's response left WPPSS right where it had been before—it could do nothing. The delay in resolving critical issues was pushing WPPSS closer and closer to default on WNP 4 and 5. WPPSS lawyers were saying that WPPSS might "default by default" because the case could not go to trial.

Anticipating the worst, WPPSS' Executive Board set aside $25 million to cover the legal and administrative costs of default. There was about $30 million in the escrow account set up by order of Judge Coleman. (If all the Participants had paid into the account, there would have been $41.3 million.) But Coleman would not release the money to WPPSS officials. Without that escrow money, WPPSS officials were saying they would be unable to make a $16 million payment into a bond reserve fund held by Chemical Bank by May 31, thus leaving WPPSS—the construction arm of the state's public utilities—in technical default. To the investment community, that missed payment into the reserve fund would be a "little 'd' default." The real crash would not come until late summer.

Missing the May 31 payment would start the clock ticking, said a Chemical Bank spokesman. If WPPSS missed the May 31 payment, Chemical Bank, trustee for the bonds, would send a notice to WPPSS on June 1 demanding payment within 90 days. If after 90 days WPPSS still had not paid, the entire $2.25 billion debt could be declared due immediately. WPPSS' financial situation then could ripen into a "big 'D' default." WNP 4 and 5 bondholders would not have to worry about not getting paid until January 1, 1984. The $94 million semiannual interest payment owed to bondholders on July 1, 1983, was already in a reserve account held at Chemical Bank. But if WPPSS did not get more money, through additional bond sales or from its Participants, it would not have the cash to put funds back in to that reserve account, and there would be no money left to pay the next semiannual interest payment, due January 1, 1984.

Chemical Bank Vice President William Berls said that the options were few if WPPSS did not meet its financial obligations. The bank might ask that a receiver be appointed, then go to court and sue for the money. The two terminated plants were worth very little, and the only assets WPPSS had were its three remaining plants, none of which were anywhere near finished. WNP 1 and 3 had been mothballed, leaving WNP 2 as the one plant most likely to be completed. If the participating utilities refused to pay even after being ordered to do so by a court, Chemical Bank could try to attach WPPSS' assets. There was also a chance that WPPSS would just limp along, three, six, even 12 months or longer. There was also the possibility of bankruptcy. But WPPSS' new Chairman of the Board, Carl Halvorson, said that the Board of Directors had recently passed a resolution saying it would never declare bankruptcy—and as a municipal corporation, WPPSS could not be forced into bankruptcy.

Eleventh Hour Attempts to Save WPPSS

As the likelihood of WPPSS' going into default on WNP 4 and 5 became more imminent, few people in Washington State worked harder to stave it off than Governor John Spellman. For months he had been working with utility and government officials, and by the end of May the strain was showing. The usually optimistic governor had grown pessimistic. "Default is in the process of occurring," he told the press. "Only some unexpected action by the 88 Participants will stop it from happening." Spellman had made very little progress during months of secret negotiating and closed-door sessions involving WPPSS Participants, business leaders, and state and local government officials looking for a solution to WPPSS' financial mess. Among the "default-avoiding" alternatives the governor and others had considered—and rejected—during months of closed sessions were:

Debt Reduction Plan. Offered by the Public Utility District Association. The 88 Participants in WNP 4 and 5 would pay $1 billion, to be used to complete WNP 3, which was about 75 percent complete and had been mothballed because of its increasing costs. (Work had been halted while WPPSS tried to secure a $960 million line of credit to ensure its completion. Standard & Poor's had suspended ratings on bonds for WNP 3, and lowered ratings made selling more revenue bonds to finance its completion very difficult, if not impossible.) WNP 1 had also been mothballed; WNP 2 was nearing completing and was not in any real jeopardy. "Since WNP 4 and 5 are lost why lose WNP 3, too?" asked the governor. Problems with the plan: Oregon and Idaho Participants would not participate, and the $1 billion could be obtained only at high interest rates.

Peak-Shaving. The debt share of the hardest hit Participants would be reduced by having other, "better off" Participants pay more than their shares. Problems with the plan: Oregon and Idaho Participants would not participate, and most of the Participants would end up paying a greater share of the WNP 4 and 5 debt.

Son of Debt Reduction Plan. WNP 4 and 5 Participants would pay $1 billion to Bonneville Power Administration, which would assume all the debt and regionalize it (that is, all electric utilities and direct service industries that received BPA power would help pay). Problems with the plan: Would impair BPA operations, the many utilities throughout the Northwest that didn't participate in WNP 4 and 5 would be asked to assume the financial burden, and the plan would require congressional approval.

Regionalization. All of the Pacific Northwest region's ratepayers (consumers of investor-owned as well as public power) would pay a portion (10 to 100 percent) of both WNP 4 and 5's debt and the debt on two investor-owned "dry holes" (Pebble Springs and

Skagit-Hanford). If all of the debt were regionalized, this plan would add 3.2 mills, or about 10 percent, to the average home customer's electric bill. (Debt obligation would vary among the utilities because they all did not buy the same percentage of BPA's electric power.) Problems with the plan: Many WNP 4 and 5 Participants would be big losers, the plan would impair BPA operations, it might be difficult to obtain congressional approval for BPA's participation, and the plan might re-ignite investor-owned power versus public power issues.

Nominal Surcharge on Innocent Parties. A surcharge would be levied on those utilities in the Pacific Northwest that had not participated on WNP 4 and 5 for "default insurance," with BPA acting as bookkeeper (levy would be one-third, one-half, or one mill). It was estimated that the surcharge would cost a typical resident an extra 72¢ a month for up to 35 years. Problem with the plan: Would be difficult to obtain congressional approval to use BPA as a bookkeeper.

All Pay or No Pay. Either all WNP 4 and 5 Participants would pay, in which case there would be no default, or no one would pay. Problem with the plan: If no one paid, the default problem would not be solved.

Cost-Sharing. Chemical Bank would drop its lawsuit against the Participants and the Participants would raise $400 million over the next 15 years, or Chemical Bank would continue its lawsuit and the Participants would raise $300 million over the next 10 years. This plan appeared to cover the Oregon-Idaho problem, provided new money for some ratepayer relief, and provided for orderly resolution of legal challenges while avoiding default. Problems with the plan: Would fail if WNP 4 and 5 Participants refused to release funds from escrow, and would not succeed if non-participating utilities had to foot all the bills to avoid default.

Hybrid Plan. Combined a millage levy on Pacific Northwest non-participants intended to give relief to Participant ratepayers with a cost-sharing settlement among 4 and 5 Participants themselves.

Governor Spellman had endorsed the cost-sharing plan, with the $300 million option and a continued Chemical Bank lawsuit. But by the end of May it was appearing that Spellman's anti-default pleas were falling on deaf ears. Jim Boldt, executive director of the Washington Public Utility District Association, said that "BPA and the investor-owned utilities still refuse to recognize we have a regional problem." The governor had even made an unusual appearance before a joint legislative session in early May in an attempt to unite the many special interests in the common cause of avoiding default. In his 15-minute televised speech, through which members of both houses sat in grim silence, he urged Washington lawmakers to pass a bill forbidding bankruptcy of WPPSS and called on the utilities to release some money to help stave off default of "this financial nightmare that is WPPSS." "Such legislation is absolutely necessary in order to ensure the fiscal stability of our region," he said of the proposed bill.

The Bankruptcy Bill

As the likelihood of WPPSS' default grew ever closer in late May, there was increasing talk of WPPSS going into voluntarily bankruptcy, even though the Board of Directors had passed a motion saying that such action never would be taken voluntarily. Governor Spellman favored bankruptcy because he, and others, believed bankruptcy would protect the assets of WNP 1, 2, and 3 from bondholder and/or creditor suits stemming from default on WNP 4 and 5. WPPSS' spokesmen said the Board of Directors could see no advantage in filing for protection under Chapter 9 of the Federal bankruptcy laws because WNP 1, 2, 3 were protected by numerous legal barriers—a virtual "Chinese Wall" of protection. "Therefore, [bankruptcy] is not in our interest nor in the interest of our bondholders," WPPSS said. Earlier, WPPSS Board Chairman Carl Halvorson had told the Seattle Society of Financial Analysts that bankruptcy was not an option and that there was little sentiment among Board members for such action. "There has been a lot of loose talk about bankruptcy," Halvorson said. "It's not a panacea." Halvorson did say that he and WPPSS would not preclude the possibility of declaring bankruptcy on WNP 4 and 5, provided that Congress modify existing bankruptcy legislation, which prevented WPPSS from going into bankruptcy on a project-by-project basis. Members of the Northwest Congressional delegation told Halvorson that while they agreed that something must be done, the rest of Congress was not sympathetic to any kind of bailout because ratepayers in the Pacific Northwest already enjoyed federally subsidized electricity from BPA's hydroelectric system.

In the meantime, a bill designed to permit limited bankruptcy for WNP 4 and 5 had been approved by Washington State

Senate Energy and Utilities Committee. However, the WPPSS bankruptcy bill was never passed by the Washington State Legislature.

With the adjournment of the Washington State Legislature's special session in late May, Governor Spellman's efforts to head off WPPSS' default came to a standstill. For Spellman, WPPSS' default represented a failure to honor obligations. "This region, this state, our people have not had a record of failure," he said. "We must not fail now." Throughout his closed-session meetings with WPPSS officials, WPPSS Participants, and the investment banking community, Spellman had searched for a settlement that would avoid default. Eight plans with dozens of variations had been considered. However, efforts had been impeded by a series of lawsuits in which courts had ruled that the Participants did not have to pay their contractual obligations. Spellman's search for a solution had also been hampered by a host of lawyers advising their clients, the Participants, not to honor their obligations unless the high courts of their states ordered them to pay. While Spellman believed that there was no long-term solution at that time, he felt that at least a short-term answer was possible, if the utilities would only agree to settle the dispute over cost-sharing. Spellman wanted a settlement whereby the utilities would put up $300 million over 10 years, with $50 million payable in the first year. The money would have been combined with funds some of the Participants had put into the escrow account ordered by King County Superior Court Judge H. Joseph Coleman. But the Participants had refused to allow any of their funds to be paid out of escrow to avoid default and Judge Coleman agreed. "Unless those payments are made, default will occur as surely as I am standing here," Spellman told the Washington State legislators. Many legislators were unmoved. Hadn't WPPSS cried "wolf" before? How could anyone believe WPPSS anyway, after its mismanagement of its entire nuclear program? Many believed that WPPSS officials were only crying "wolf" again and that their word was not to be trusted.

Washington Supreme Court Rules Utilities Don't Have to Pay

On June 15, 1983, people holding WNP 4 and 5 bonds came closer than ever to losing their investment. In a 7–2 decision, the Washington State Supreme Court overturned key elements of King

County Superior Court Judge Coleman's November 1982 opinion, ruling that Washington public utilities had not had authority to sign contracts to participate in WNP 4 and 5. The high court held that the utilities, including 19 PUDs, an irrigation district, and 13 electric cooperatives in the state, had had neither express nor implied legal authority to enter into agreements to pay for the plants being built by WPPSS. The Participants' Agreements were illegal and thus not enforceable, the court majority said in its opinion by Justice Robert F. Brachtenbach. Coleman had also ruled that the utilities were obligated to pay whether the plants were furnished or not but the Supreme Court did not rule on this so-called "hell or high water" provision.

The Washington Supreme Court said the utilities, as creatures of Washington State, had never been granted authority by the legislature to enter into contracts allowing them to share in "capability"; existing law allowed them only to contract to buy or sell power, the court majority said. Nor did the law allowing the utilities to build and operate their own generating facilities allow such contracts, the majority said. According to Justice Brachtenbach:

> In the present case, the participants lacked substantive authority to enter into this type of contract because they constructed an elaborate financing arrangement that required the participants to guarantee bond payments irrespective of whether the plant was ever completed; to surrender ownership interest and considerable control to WPPSS (Washington Public Power Supply System); and to assume the obligations of defaulting participants. As such these contracts failed to protect the unsuspecting individuals, the ratepayers, represented by the participants. By choosing such an alternative arrangement in lieu of a statutory scheme that incorporated protections against these very liabilities, the participants exceeded their statutory authority, rendering the contracts ultra vires [outside the law].
>
> Although we have discussed our conclusions throughout the opinion, it is helpful to summarize. It should be apparent that this agreement does not satisfy the statutory scheme governing the public participants:
>
> (1) The agreement is not a standard contract for the purchase of power because the payments are due irrespective of whether any electric current is delivered.

> (2) It is not the type of acquisition or construction of a generating project authorized by the statutes or previously recognized by this court, because the participants retained no ownership interest, except in any excess assets upon termination, and a very limited role in management of the project.
>
> (3) It is not an exercise of an implied power to pay for municipal services because there was no guaranty the services would be provided and we perceive no legal necessity for such powers.
>
> (4) Finally, it is not just a joint operating agreement with the provisions of RCW [Revised Code of Washington] 43.52 [the law authorizing WPPSS joint operating agreements] because those provisions limit the Participants' ability to buy anything more than electric energy.
>
> Our conclusion is that the Washington statutes authorize the Participants to purchase power or to own electric generating facilities. An attempt to structure an agreement under the joint operating agency statutes or to base it upon implied powers does not alter that basic authority and the participants simply are not authorized to guarantee another party's ownership of a generating facility in exchange for a possible share of any electricity generated. Therefore, we hold that the Washington PUDs and Washington municipal participants lacked authority to enter into this agreement.

Justice James M. Dolliver joined Justice Robert F. Utter in his dissent, which read in part:

> The majority, by its narrow reading of municipal authority to provide electric power, places constraints on municipalities not intended by the Legislature. Municipal authority in this area must be read broadly to provide the flexibility which is absolutely crucial in furnishing such a capital-intensive service. The contractual agreement in the present case is a form of purchase which, viewed at the time the contract was made, may well have been the most economically advantageous for all concerned. While the result has been tragic, the decision here today may not free municipalities of all potential costs in even the present case, will bar them from potentially advantageous contracts in the future, and may well make financing of future projects more costly.

> It is natural for anyone viewing the enormity of this mismanagement in this project and the calamitous impact of its failure on utilities and ratepayers to seek ways to negate the impact. Nonetheless, I cannot agree that the approach taken by the majority is either the proper or necessary way to solve the problem.
>
> The issue before us whether a proper construction of this language [allowing utilities to buy power] includes the purchase of a possibility of electric power. The majority simply concludes that it does not. I find the issue more difficult.
>
> I believe this court should construe [the law] to authorize purchase of a possibility of power as well as more typical purchase contracts. Such contracts in some circumstances may be highly advisable. Option contracts, especially when used in an unpredictable market in which periodic gluts and shortages are common, are a good example.
>
> Several years ago, for example, a municipality operating an oil-burning plant seemingly might have been well advised, in the face of escalating oil prices, to enter a long-term contract to purchase oil at the then-prevailing price. Yet in light of recent price declines, such a contract would have left the municipality bound to pay significantly more than current price. The solution—an option contract which would have permitted the municipality, by exercising its option, to avoid any price increases and yet still, by choosing not to exercise its option, take advantage of any price declines.
>
> The participants are no more entitled to a refund of their consideration here, that is, relief from their bond guaranties, than they would be entitled to a refund of consideration paid for the option contract.

Cheers and Dire Warnings

Those who thought the Participants should not have to pay off the debt on $2.25 billion WNP 4 and 5 bond issue naturally were delighted with the State Supreme Court's ruling. Carol Dobyns, president of The Light Brigade, an anti-WPPSS ratepayer group, called the ruling "wonderful." "We always knew that the ratepayers were not responsible for the bill for this mess," she said, "We never, ever wanted to make a deal with those people. We wanted to fight them, and we won." Lawyers for the utilities said they were not surprised by the decision. Albert Malanca, a Tacoma lawyer representing a group of utilities, said the court "clearly ruled" that the utilities would not have to pay. "For [public utility district] commissioners to stand up now and say they would pay—they would be run out of town on a rail."

WPPSS officials were shaken by the turn of events. "This is a devastating decision," said Don Mazur, the new WPPSS Managing Director. Members of the WPPSS Executive Board discussed holding an emergency meeting, "but I don't know what the hell we can do," said Carl Halvorson, Executive Board Chairman. "This is going to hurt everywhere, not just us, but everyone." Analysts had warned that a WPPSS default could severely harm other Pacific Northwest municipalities and agencies in their future efforts to sell bonds. On the day the Washington Supreme Court's decision was announced, prices of other revenue bonds in the general market for tax-exempt securities "knee-jerked" down as much as a point initially, though but they did come back rather quickly. (The movement of a point is equivalent to a $10 change in the price of a bond with a $1,000 face value.) Trading in WPPSS bonds dried up following the court's action, with traders shunning the name until the smoke cleared a bit. Investors felt that there wasn't any lasting damage to the revenue bond market. The damage appeared to be confined to the WPPSS name: WNP 1, 2, and 3 bonds fell about four points on the day of the court's decision and WNP 4 and 5 bonds, more than 15.

Chemical Bank Asks Court to Reconsider

Despite the Washington State Supreme Court's decision, investment bankers and others cautioned that it would not be the end of court action. On July 5, Chemical Bank asked Washington State's highest court to reconsider its "legally indefensible" decision that freed numerous utilities from paying for their part in WNP 4 and 5. In a sharply worded, 54-page document filed with the court, attorneys went on to say that the decision was "neither wise nor correct and ... should be reversed upon reconsideration." In addition, the attorneys had this to say about the 7–2 decision:

> ... the majority ignored state law that gives public utilities clear authority—both expressed and implied—to participate in a power project, even when there is a risk it will turn out to be a "dry hole."

> Because of the magnitude of the decision and the matters overlooked by the court, additional consideration and argument is clearly warranted.
>
> Reconsideration will show that the decision contravenes express statutory language, clearly stated intent, the law of implied municipal powers, and constitutional provisions requiring due process of law and prohibiting impairment of contracts.
>
> By undermining the power of public utilities, the court threatens to erode the fundamental principle of separation of powers and of public faith in the certainty and stability of legal rules.
>
> The court majority has strained to interpret the provisions [on the utilities' authority] narrowly. Contracts signed by the utilities are valid and authorized undertakings of the participants.
>
> The utilities had full control over WPPSS and the WNP 4 and 5 and were not the unwitting victims of some outside agency.
>
> The decision of this court obviously has a profound impact on bondholder's rights, in direct violation of the constitutional guarantee that contracts will not be abridged.
>
> When times were good, the utilities signed up for the plants. When the circumstances changed in January 1982, most of those municipalities immediately repudiated their promises and groped for a justification that is so unprecedented that virtually none of the municipalities, who asserted seemingly innumerable frivolous grounds for evading their promises, even attempted to argue it.
>
> While the majority's opinion is no doubt politically popular among some people in this region, it is legally indefensible, and it obliterates the fundamental distinction between the legislative and judicial functions. The court's concern for those "unsuspecting individuals, the ratepayers," is misplaced. They had the responsibility to elect qualified utility managers and commissioners and the right to make those officials answerable to their electorate. Those ratepayers had the opportunity to attend public meetings and participate in every phase of the process that has resulted in this tragic financial loss. The true victims in this case are the bondholders....

A Supreme Court reconsideration is a pretty rare event. Rehearings almost never are granted unless the decision was a 5–4 split or there has been a turnover of justices on the bench. Although asking the Washington Supreme Court to reconsider might seem to be an exercise in futility, it was felt to be the last chance Chemical

Bank had to make the utilities pay. The bank and its attorneys figured they might be in for a long wait while the court decided whether to grant a rehearing. In the meantime they began studying the WPPSS case for "possible Federal issues" that might allow an appeal to the U.S. Supreme Court, though they aknowledged that there did not appear to be a clear-cut Federal issue in the decision.

Chemical Bank did not have to wait long to hear from the Washington Supreme Court. On July 22, 1983, Justice William Williams signed a brief order denying the motion for a rehearing of its previous ruling, saying only that the decision to deny a rehearing had been made by a majority vote of the nine-member court.

If 46 Don't Pay, Neither Do the Other 42

The earlier rulings had not applied to the 42 rural electric cooperatives located in Washington, Oregon, Idaho, Montana, Nevada, and Wyoming which operate under Federal authority. Since all 88 Participants had been named in Chemical Bank's summer 1982 suit and that suit was still pending, there were loose ends to tie up. Attorneys for the co-ops asked Judge Coleman to dismiss the Chemical Bank suit, contending that the bank's claim had been so seriously damaged by the Washington Supreme Court's decision, which affected about 70 percent of WNP 4 and 5's ownership, that their clients should also be freed from their obligations to pay. The attorneys also pointed out that virtually every rural electric co-op involved with WPPSS would be forced out of existence if forced to pay the $2.25 billion debt without help from the large public utility districts and municipalities. Judge Coleman basically agreed with that argument and on August 10, 1983, ruled that the 42 rural electric cooperatives participating in WNP 4 and 5 did not have to pay their share of the $2.25 billion bond debt, thereby completing the job begun in the Oregon courts. Now all 88 Participants had been absolved of their debt on WNP 4 and 5.

Contracts Between Agencies and Participants: Bond Buyer Beware!

Many people have questioned the justice of the Washington Supreme Court's decision to free dozens of public utility districts and municipalities from contractual obligations to pay WNP 4 and 5's debt. The court based its 7–2 decision on the premise that the Participants had not had the authority to enter into contracts containing a "dry hole provision" that required them to pay off the bonds whether or not any usable power was ever generated by the plants. The issue was whether a Washington municipal corporation had the authority unconditionally to guarantee revenue bonds issued by a joint agency like WPPSS in return for the joint agency's nonguaranteed delivery of electric power. "As a general rule," wrote Washington Supreme Court Justice Robert Brachtenbach for the majority in *Chemical Bank v. WPPSS, No. 49186-1*, "the unauthorized contracts of government entities are rendered void and unenforceable under the ultra vires doctrine." The "ultra vires" (outside the law) doctrine is an established principle of municipal law and exists to "protect the citizens and taxpayers ... from unjust, ill-considered, or extortionate contracts." The high court's conclusion was, in short, that the "dry hole" provision in the Participants' Agreements the WNP 4 and 5 Participants had signed had never been authorized, either expressly or implied, by Washington State statute, and that therefore the Participants' Agreements were void and unenforceable. The State of Washington said via *Chemical Bank v. WPPSS* that a municipal corporation had no such authority. Courts in Texas, Louisiana, South Carolina, Georgia, and Wyoming have disagreed, with various qualifications, and at least five other states – Maine, Massachusetts, North Carolina, Vermont, and Virginia – have statutes indicating that a municipal corporation does have that authority.

In reaching its decision, the Washington Supreme Court distinguished between the explicit authority of the utilities to purchase electricity on behalf of their customers and the agreement in question, which involved the purchase of "project capability." Clearly the WNP 4 and 5 Participants had statutory authority, under RCW (Revised Codes of Washington) §54.16.040, to purchase electricity. Project capability included the possibility that no power would ever be generated, and, the Court said, "the participants could [wind up paying] approximately $7 billion for nuclear plants which would never generate any electricity." Ultimately the rate-paying consumers would pay for the nonexistent electricity. "The unconditional obligation to pay for no electricity is hardly the purchase of electricity," Brachtenbach wrote. "[E]ven the express power to buy and sell electricity or to acquire or construct generating facilities may not be exercised in a manner beyond the scope of that primary purpose."

The Washington Supreme Court also examined the question of whether the express authority to pursue means of bringing power to people carried with it any implied powers. The majority of the Justices recognized that the bond guarantees may have been necessary to sell the bonds in the investment market, but said: "Necessity does not provide authority. Accordingly, we do not believe that [the Participants' Agreement] is authorized as an implied power to pay for an admittedly proper municipal service." The Court also ruled on the provision in the Participants' Agreement that if any Participant to the agreement defaulted, the remaining Participants would be required to make up any deficiencies in funds. The Court observed that this contractual agreement was contrary to the statutory requirement that each party be responsible only for its own debts and obligations in a facility.

Adjucating under a strict constructionist philosophy, the Court did not find an implied statutory authority. It held that a municipal corporation's powers are limited to those conferred in express terms or those necessarily implied, and that the test for necessity is legal, not practical. The Court noted that although in some states "home rule" powers of municipalities result in considerable autonomy from state control, Washington courts have interpreted the home rule powers of first-class cities more narrowly.

The Court might have ruled differently had it found an ownership interest in WNP 4 and 5, which would be acceptable under the statutory powers. As part of the responsibility of ownership, utilities might have been able to issue valid guarantees for the bonds. But the Court found that the utilities maintained only a limited management role, which was not sufficient to constitute an ownership interest. Although the Court said that the existence of a "Participants' Committee, which met periodically to approve or disapprove major decisions, suggested a degree of participant control and management," it found the Committee to be nothing more than a "rubber stamp" for WPPSS' decisions. "We are not prepared," the Court said, "to sanction a virtual abdication of all management functions and policy decisions to an operating agency such as WPPSS. Here the participants' committee apparently served as 'rubber stamp' for WPPSS' decisions, resulting in two terminated projects, less than 25 percent complete, at a cost of $2.25 billion, or almost $7 billion over the 30-year repayment period." Added the Court: "As a matter of public policy, the enormous risk to ratepayers [the possibility of a dry hole] must be balanced by either the benefit of ownership or substantial management control."

A dissent by Justice Robert Utter criticized the majority's "narrow reading of municipal authority to provide electric power." He contended that the decision "places constraints on municipalities not intended by the legislature." "Municipalities exercising their authority to provide electric power to their citizens," Utter wrote, "must be given the freedom and flexibility to use all advisable means. Their determinations of advisability, moreover, are not subject to judicial review."

14

Default and Its Aftermath

On July 25, 1983, the Washington Public Power Supply System acknowledged that it could no longer pay interest and other obligations on Washington Nuclear Plants 4 and 5. WPPSS' admission had come in response to an interrogatory from Chemical Bank of New York, the bond trustee: "Does the Supply System admit its inability to pay its debts incurred in the ownership and operation of WNP 4 and 5 generally as they become due?" WPPSS Deputy Managing Director Alexander Squire had delivered a one-word answer: "Yes." Even though WPPSS had not yet missed an interest payment, William Berls, a vice president in Chemical Bank's trust division, said Squire's admission constituted an "event of default," permitting the bank to take action on behalf of the estimated 75,000 to 100,000 WNP 4 and 5 bondholders across the country.

As trustee, Chemical Bank's job was (and is) to act on behalf of the bondholders; the bank itself is not liable for the repayment of the debt. Chemical's alternatives included demanding immediate repayment of bonds sold to finance 4 and 5 and the interest on them, which action would strengthen its position in any future lawsuits. Another alternative was move in and take over the assets of WNP 4 and 5, acting on its own or through an agent or receiver. A third option was to sell the equipment, materials, and supplies from the two partially built nuclear reactors. Certainly, WPPSS had a great many items to sell. During the mid-1970s, WPPSS had tried to plan ahead and beat skyrocketing inflation by purchasing 90 percent of everything it needed to operate the two nuclear units, much of it even before construction began. (Some of the equipment, bought and paid for, is not even at the construction site. For example, two steam turbine generators, each costing $44 million, have never left their manufacturing site in Lancaster, Pennsylvania. In 1983, WPPSS paid $44,600 just to store them.)

Right after WNP 4 and 5 were terminated, WPPSS officials went looking for someone to buy the whole package and maybe possibly complete the nuclear project. They couldn't find a suitable buyer, so they began selling off whatever they could, piece by piece. Money from the "disposition of assets," an estimated one to five cents on every dollar of the $2.25 billion spent, would be turned over to Chemical Bank, since it in effect, was now the owner of the two "dry holes." Buyers could get a good deal on the two sets of nuclear reactor vessels: 50 feet high and 15 feet wide and made of a special, high-grade steel nearly 10 inches thick, the original price for each set of reactors was $204 million. But because no new nuclear plants are planned for construction in the United States, both sets of reactor vessels will probably wind up as scrap metal. Additional items for sale include 1,000 miles of unused electrical cable, more than 100 miles of pipe, and thousands of tons of steel reinforcing bar. But most of the parts that are up for sale were custom-designed for WPPSS and would not be usable on other projects. In addition, many of the parts are already in place, and equipment is much less salable when it's installed. Some of the items would be too large to ship anywhere else. For example, there were four steam generators, one pair weighing 540 tons each and the other 740 tons each. WPPSS had gone to great lengths to get these generators to the construction sites from Chattanooga, Tennessee, where they were built. From Chattanooga they were shipped down the Mississippi River, into the Gulf of Mexico, through the Panama Canal and all the way up the Pacific coast to Washington State. WPPSS had even had to build a $8.3 million road to support the transport of this and other heavy equipment.

Bail-Out Efforts

Even before WPPSS acknowledged in July 1983 that it could not pay its debts, nearly everyone involved—WPPSS management, construction unions, bondholders, public utility districts, investment bankers, state and local officials—had hoped to interest the Federal government in a bail-out action. Most of the ideas had been around as early as 1982, when Washington Governor John Spellman had hired Shearson/American Express to develop a Federal "rescue" plan. This and other plans had received very little attention in Congress, mainly because beleaguered Federal taxpayers in the recession years of the early 1980s were in no mood for a multi-billion dollar bail-out necessitated by alleged mismanagement. It didn't help the cause any when Congress found that ratepayers in the Pacific Northwest region had the lowest electrical rates in the United

States, thanks to federally built hydroelectric dams. Then, too, WPPSS had not yet actually defaulted, and some wondered if officials were merely crying "wolf." Although WPPSS' acknowledgment that it would have to default did not come as a surprise to members of the U.S. Congress, it did strengthen somewhat the belief among many Pacific Northwest congressmen that a Federal rescue action of some sort should at least be given a try. Senator James A. McClure (R–Idaho) introduced in Congress in late July 1983 a WPPSS rescue measure that would have provided fresh funding for completion of WNP 3. Certainly, the measure was not comprehensive, but it was a start.

WNP 3 had been mothballed in July 1983, just a few months after WNP 1 suffered that fate, because no one was willing to lend WPPSS any more money. Unless some way could be found to finance their completion, WNP 1 and 3 also would be terminated, becoming "dry holes" into which $5.1 billion had already been poured. WPPSS was still planning to complete WNP 2, but only because Bonneville Power Administration had agreed to provide a loan of $150 million or more to finish it. Few people had complained when WNP 1 was mothballed. However, with the mothballing of WNP 3, which was 75 percent complete, and talk of even terminating it, WPPSS angered the four investor-owned utilities that together owned a 30 percent share in the nuclear plant—Pacific Power & Light Company, Portland General Electric Company, Puget Sound Power & Light Company, and Washington Water Power. WPPSS officials felt that they had had to mothball the plant because of WPPSS' serious financial problems. But the Board of Directors had made the decision unilaterally, disregarding the fact that the four investor-owned utilities had put about $1 billion of their own money into the nuclear project. Seeking to protect their investment, three of the four utilities on July 29, 1983, filed an "amended answer" to a previously filed suit in U.S. District Court, using this legal vehicle to assert their claims arising from the mothballing, and possible later termination, of WNP 3. (Pacific Power & Light Company, which owned 10 percent of WNP 3, did not join the complaint.) The three investor-owned utilities threatened to sue if work was not resumed. They charged that neither BPA or WPPSS could legally take action that ran counter to their rights under their WNP 3 ownership agreement. According to the three utilities, WPPSS was obligated to complete and pay for its share of construction of WNP 3 and to have the nuclear plant operational by December 1986, as previously agreed.

To help ensure that WNP 3 would be completed as agreed, the four investor-owned utilities privately sought out friendly congressmen to help them secure credit from two New York City banks. The banks, Chase Manhattan and Citicorp reportedly, had told the utilities, WPPSS, and BPA officials that they would grant WPPSS a $960 million line of credit if an independent financing corporation were established and BPA guaranteed that the money would be repaid. BPA was willing to go along with these conditions, but it could do so only if it received congressional authority to make such loan guarantees.

Making the Best of a Bad Situation

Sensing that WPPSS' entire nuclear power plant program might be on the verge of a financial and litigation meltdown, one of the supporters of public power in Congress, Senator James A. McClure (R–Idaho), chairman of a subcommittee of the powerful Senate Appropriations Committee, listened with great interest to the lobbyists from the Pacific Northwest. Perhaps the most serious effect of a Northwest entangled in decade of WPPSS litigation, the Senator was told, was the millions of additional dollars in interest public entities would have to pay when they went to the bond market. A further disheartening development of the WPPSS default was that management personnel at the utilities were being distracted from their main mission of providing safe, economical, and plentiful energy. Managerial decision-making was now being passed from professional energy managers to lawyers, and, not unnaturally, McClure was told, the lawyers were more concerned with planning the best litigation strategy than with planning for the region's energy future. If the necessary funds could not be obtained to complete WNP 3, other generating facilities fueled by other sources, presumably coal, eventually would have to be substituted – at a cost far higher than the cost of completing WNP 3, and WNP 1 and 2 as well.

But what really concerned the officials from the Pacific Northwest was what they described as the "WPPSS Legacy," the "costs" of the WPPSS experience that went beyond the immediate situation. For one thing, the market price for WNP 4 and 5 bonds had fallen to a small fraction of their original cost. (By November 1983 WNP 4 and 5 bonds were selling at discounts approximately 85 full percentage points below their face value.) Even the bonds issued to finance WNP 1, 2, and 3—bonds indirectly backed by the credit of BPA—had been downgraded from AAA to A+, then suspended entirely. (As of November 1983, WNP 1, 2, and 3 bonds were selling at discounts of 30 or more full percentage points below the price of AAA bonds having similar coupon rates and maturities.)

Moreover, the impact of WPPSS' problems was extending beyond WPPSS itself. Immediately after WPPSS' event of default, the four investor-owned utilities involved in WNP 3 were placed on a "credit watch" by Standard & Poor's (they were later removed from that list). And the cost of default was already showing up in the bond market for funds for other civil projects. For example, the Snohomish County Public Utility District, which owned the largest share of WNP 4 and 5 (13 percent) and had led the fight to avoid paying its share of the $2.25 billion bonded debt, had had its Moody's overall bond rating downgraded to A from A1 and its Standard & Poor's overall rating downgraded to A from A+. (Snohomish PUD's November 14, 1983, issue of $243 million of revenue bonds to finance completion of its Sultan hydroelectric project was rated by Moody's as Baa1 and by Standard & Poor's as BBB+. Although the issue was sold out, it was at 130 basis points—or 1.3%—above the market for an A rated revenue bond; this added a total *extra* cost of $122 million over the 30-year life of the bonds.)

Another example of the negative impact of the "WPPSS Legacy" was the State of Washington's issue of $150 million of general obligation bonds in August 1983, apparently at 76 basis points—about three-quarters of a percent—above the market for that type of bond—adding a total *extra* cost of $18 million over the 25-year life of the bonds. (In a later sale of $131 million in bonds, Washington apparently paid a penalty of 55 basis points—a total *extra* cost of almost $12 million.) Clearly, investors were demanding higher interest rates on securities offered by borrowers in the Northwest. For example, one small block of bonds issued by the Tacoma, Washington, electric revenue authority sold in May 1983 with an 11 percent yield, whereas for similar bonds issued by authorities in other parts of the country sold with a yield of slightly less than 10 percent.

The Pacific Northwest officials lobbying Congress also pointed out that the "WPPSS Legacy" wasn't confined to the Pacific Northwest region. Outside the Pacific Northwest, public power agencies that financed their projects by selling revenue bonds were beginning to believe that the WPPSS default was costing them, too—as much as 50 or more basis points on new issues. Among the revenue bond issues for electricity-generating projects hardest hit by WPPSS' problems was the Massachusetts Municipal Wholesale Electric Cooperative, which had postponed a large bond sale after WPPSS' default. "Massachusetts Wholesale does have a financing problem, no doubt about it," said Robert Adler, head of municipal bond research at Shearson/American Express. Although Massachusetts communities and local utilities had not refused to make payments, as had those in the Northwest, investors were alarmed at the huge costs of completing the Seabrook (New Hampshire) Nuclear Project, in which the Massachusetts authority has a big interest. Indicative of investor concern, the 10.125 percent Massachusetts bonds due in 2017 were offered in mid-July 1983 at a price boosted by 82 basis points to yield 12.89 percent, well above the 9.9 percent return for many other non-nuclear electric revenue issues.

Another new issue known to have been withdrawn from the bond market in mid-1983 because of the adverse impact of the WPPSS default, the officials told McClure, was a $349 million North Carolina Municipal Power Agency (NCMPA) issue which was based on a take-or-pay contract with local utilities similar to WPPSS' contract. (After withdrawing its $349 million issue, NCMPA stated its intention to seek congressional action to undo the damage it believed the WPPSS default had caused. After the WPPSS 4 and 5 default, all municipal electric revenue bonds had become suspect, NCMPA said, and it was having to pay one-half to one percent more on its bonds than it would have without WPPSS. By November 1983 NCMPA was circulating among other publicly owned joint utility operating agencies a draft of a proposed congressional bill that would require BPA to pay off the WNP 4 and 5 bonds. Called the "Pacific Northwest Regional Payment Act," the bill would authorize and direct the administrator of BPA to make payments to investors in bonds issued by WPPSS in connection with its ownership share of WNP 4 and 5, from the revenues received by BPA from its electric customers.)

Impressed by the seriousness of what he heard, Idaho Senator McClure acted almost immediately by attaching a "rider" to the pending $7.6 billion Interior Department appropriations bill in late July 1983 that would have helped the owners of WNP 3 obtain fresh funding from New York City banks for completion of that plant. McClure's rider had the support of Oregon Senator Mark Hatfield, Washington Senator Henry Jackson, Washington Governor John Spellman, and Idaho Governor John Evans. Approval of McClure's proposed rider by the Senate Appropriations Committee was virtually assured, for that committee was chaired by Senator Hatfield. Under the provisions of McClure's rider, known as Senate Bill S.1701 (the bill was identical to section 317 of H.R. 3363, the 1984 appropriation bill for the Interior Department and related agencies), BPA would be empowered to enter into alternative financing agreements through an appropriate agency set up in accordance with Washington State law. The new agency would issue bonds or other types of indebtedness. The security for lenders would be the contract established between the new financing agency and BPA. This contract would commit BPA to pay the principal, interest, and related costs on the new borrowings directly to the new financing agency or its trustees. On July 19, McClure's rider was approved on a voice vote and sent to the Senate floor, where it would be considered by the entire Senate. "Since all the [Pacific] Northwest are for this, I expect it will have no trouble going through the Senate," said House Appropriations Committee member Norm Dicks (D–Wash), "but the bill will obviously be much more controversial when it reaches the House."

Even before Senator McClure's proposed rider to the upcoming Interior Department's appropriation bill was approved by the Senate Appropriations Committee in July 1983, there was evidence that such a measure would be strongly opposed. Oregon U.S. Representative James Weaver, who chaired a House subcommittee overseeing BPA, had denounced McClure's WPPSS rescue plan as "a blank check for BPA" and had said he would "fight it in every way possible." "The stench over WPPSS is growing worse," said Weaver. "Electric rates could go up as much as $1.15 billion to complete this plant (WNP 3)." BPA spokesman Ed Mosey was cool to Senator McClure's proposal, too:

> It's the first living, breathing bill before Congress that deals with the Supply System. However, it does not address the question of [WPPSS'] threatened bankruptcy. It doesn't address whether plant financing would be prudent. Nonetheless, it is a major first step.

First step or not, opponents of McClure's proposal were convinced that the idea was badly flawed. Republican Representative Rod Chandler of Tacoma, Washington, said he would refuse to support any financial assistance plan for WPPSS that did not include paying off the debts on WNP 4 and 5. Said Democratic Representative Al Swift of Washington, a member of the House Committee on Energy subcommittee on energy conservation and power: "Judgments were made by people who don't know a tinker's damn about legislation who bulled ahead with this." Democratic Representative John D. Dingell of Michigan called the Senate measure "rather slipshod legislation ... richly surrounded by a massive growth of questions." Representative Richard L. Ottinger (D–New York) said: "[C]oncocted behind closed doors by four private utilities owning 30 percent of the project, the provision creates more problems than it solves, including a potential of complete liability by the taxpayers." Dingell and Ottinger called McClure's proposal "a blank check in the hands of the BPA administrator. It could cost BPA ratepayers dearly and there is no evidence that such unlimited authority is needed." In addition, the two congressmen said it was not clear that paying for nuclear plant construction costs through a new financing entity would make such payments any more prudent. "This provision should be rejected," Dingell and Ottinger wrote in a letter to Illinois Representative Sidney Yates, chairman of the House Appropriations Interior Subcommittee, which would participate in the House–Senate conference that would wrestle with McClure's rider if and when it cleared the full Senate. "If legislation is needed, let BPA and the private utilities inform appropriate legislative committees of the House and Senate in an open manner," they wrote.

On August 4, McClure's increasingly controversial WPPSS rider to the Interior Department appropriations bill ran into a "mini-filibuster" led by Senators William Proxmire (D–Wisconsin) and Howard Metzenbaum (D–Ohio) before other Senate business ended the opposition's debate. Metzenbaum opposed McClure's rescue plan because of BPA's involvement in it. In addition, he noted that BPA had failed to make payments on its longstanding $7.9 billion debt to the Federal treasury – a debt that BPA appeared in no hurry to repay. If McClure's rider passed, Metzenbaum was ready to introduce legislation forcing BPA into a rigid repayment plan. With this pressure, McClure agreed to drop his plan, for if Metzenbaum's threatened bill would have added an estimated $9 billion to power rates in the Pacific Northwest region over the next 10 to 20 years. With McClure's rider withdrawn, the Senate adjourned for its summer (1983) recess without having to vote on McClure's proposed WPPSS rescue plan. McClure promised he'd try again in the fall to salvage $1 billion or more for WNP 3. He charged that his rescue effort had been defeated by environmentalists "who want to kill those plants." "There is a linkage of people, who, for other reasons, want to kill those [WPPSS] nuclear plants ... the people from the environmental coalitions who infest this town [Washington, D.C.]," he said.

The Luce Commission's Report

In late summer 1983, Governors John Spellman of Washington and Victor Atiyeh of Oregon established a three-member advisory panel to investigate and "recommend solutions to the major problems involving, or arising from, the difficulties associated with the Washington Public Power Supply System." (According to Governor Spellman, the action had the blessings of Idaho Governor John Evans and Montana Governor Ted Schwinden.) This was the second blue-ribbon panel formed by Spellman and Atiyeh to study WPPSS problems—and the third attempt by Spellman to halt the financial meltdown of WPPSS. This third effort drew some criticism, especially from groups that thought the panel should include customers and public power representatives, not just prominent businessmen. U.S. Representative Mike Lowry (D–Washington) said that the panel members would be perceived "as the sort of people who got us into WPPSS in the first place" and that the panel's recommendations would carry little weight. Spellman defended his actions, saying "I am the only one who has taken any initiatives on this." "My feeling is that those who criticize have never taken any steps to solve this crisis themselves." The members of this second blue-ribbon panel were Edward E. Carlson (also a member of the first blue-ribbon panel), Charles F. Luce, a Portland attorney and retired Consolidated Edison Chairman (also a member of the first blue-ribbon panel), and Herbert M. Schwab, a retired Oregon Superior Court Judge. (Luce had also been an Administrator of the Bonneville Power Administration, a member of the WPPSS Executive Committee, and a $200 per hour consultant to WPPSS working to find electric power markets in California that would allow completion of WNP 4 and 5.)

The panel was asked to seek solutions to the problems the Pacific Northwest faced, not to fix blame for what had gone wrong—to focus on the future, not on the past. The panelists were asked to keep three objectives uppermost in their minds:

1. In the interest of consumers, maintain the lowest practicable electric rates consistent with sound business practice;

2. In the interest of the Pacific Northwest, maintain a reliable power supply to support economic growth and to increase job opportunities; and
3. In the interest of all states in the region and the national interest, protect the credit of state and municipal agencies and their access to capital markets to finance needed public improvements.

The panel was assisted in its technical work by BPA and the Pacific Northwest Utilities Conference Committee (PNUCC). (It was PNUCC's high energy demand forecasts in the early 1970s, which later proved vastly erroneous, that had contributed to the decision to build WNP 4 and 5.) The panel's legal work was handled by the Portland, Oregon, law firm of Preston, Thorgrimson Ellis & Holman, in which Luce was a partner. (The firm's partners also included one former congressman and several former Senate aides.)

This panel was formed about the same time the Congress was considering Senator McClure's "WPPSS rescue" plan. Spellman had endorsed McClure's plan without waiting to see what the panel proposed. As it turned out, several of its recommendations were similar to what the investor-owned utilities had urged Pacific Northwest senators such as McClure, Jackson, and Hatfield to push for in Congress. However, the "Luce Report," which was released in November 1983, went further and offered a far more detailed plan for solving WPPSS' problems.

First, the blue-ribbon panel recommended that the mothballed WNP 1 and 3 be completed. Second, the panel said that a compromise settlement with the investors in WNP 4 and 5 was highly desirable and would be much less costly to the ratepayers of the region than bitter and protracted litigation. The panel members suggested that "out of fairness as well as practicality" the cost of a settlement should be shared on a region-wide basis. The settlement they suggested would cost the average ratepayer in the Pacific Northwest, who used an average of 1,200 kilowatt hours per month, approximately 72¢ per month. They noted that years of lawsuits over the rights of WNP 4 and 5 bondholders could make it impossible to finance completion of WNP 1 and 3, in which the ratepayers had already invested more than $5 billion. Moreover it was conceivable that WNP 4 and 5 bondholders might win the lawsuits. Legal expenses alone would cost ratepayers "tens of millions of dollars." On top of all those direct costs, the panel noted, the litigation alternatives would impose substantial indirect costs on all types of public agencies throughout the Pacific Northwest by damaging the credit of the region.

To carry out its recommendations, the panel suggested that a new Federal corporation, with regional directors, be established to take over the nuclear program from WPPSS. The corporation would assume both the assets and liabilities of WPPSS. It would also assume the defense of, and seek to settle, the litigation against WPPSS, the 23 public utilities represented on WPPSS' Board of Directors, the 88 Participants in WNP 4 and 5, and other members of WPPSS' governing bodies who had been sued by bondholders. The new corporation would pursue vigorously, they said, all claims against third parties who may have had responsibility for the losses associated with the WPPSS nuclear construction program.

> There are, of course, parts of our recommendations with which particular organizations can be expected to disagree, and the Congress and the legislatures may well decide to make changes in the recommendations. We believe, however, that our recommendations offer a broad framework on which to construct a comprehensive solution to the many WPPSS related problems.

Post-Default Litigation

The legal suits spawned by the Washington State Supreme Court's release of the Washington Participants from WNP 4 and 5's $2.25 billion debt and the subsequent WPPSS default have grown numerous and complex. Claims now running into the billions of dollars have been filed in various Federal courts, including the U.S. Court of Claims, and in the courts of five states. Suits on behalf of the 100,000 plus present and former WNP 4 and bondholders have been brought in assorted class actions, and by the Chemical Bank of New York as trustee for present bondholders. Inevitably these suits will involve the thousands of WNP 1, 2, and 3 bondholders, too, as the WNP 4 and 5 bondholders assert claims against all the assets of WPPSS and its members. Various utilities are suing each other. PUD commissioners, city councilmen, BPA, consulting engineers, underwriters, lawyers—all are caught in a net of litigation that may take 10 years or more to untangle, and the results are virtually impossible to predict.

Within less than two months after the Washington State Supreme Court held invalid most of the Participants' "hell or high water" contracts that provided the security for the WNP 4 and 5 bonds, Chemical Bank filed suit in the U.S. District Court for the Western District of Washington seeking damages and other relief from the 88 Participants and others. The suit is based primarily on allegations of fraud, misrepresentation, violation of Federal and state securities laws, negligence, and unjust enrichment. It joined as defendants some 630 individuals and municipal and private corporations—not only WPPSS and 87 Participants (Orcas Power & Light, a small cooperative, then in bankruptcy, was not joined), but also all public agencies represented on the WPPSS Board, BPA, and the individuals who were serving on, or had served on, the boards of WPPSS, and the members of WPPSS. For good measure, the Chemical Bank suit includes 100 "John Doe" defendants.

In addition to the Chemical Bank's omnibus suit, a number of class actions have been filed on behalf of present and former bondholders, seeking damages not only from the defendants named in the Chemical Bank suit, but also from bond underwriters, credit rating agencies, and the many lawyers who had provided legal opinions that the Participants had authority to enter into the security arrangements underlying the WNP 4 and 5 bonds.

In November 1983, the Securities and Exchange Commission (SEC) admitted that it was investigating a number of possible culprits in the massive July 1983 default on bonds issued by the Washington Public Power Supply System. SEC investigators were searching for evidence of securities fraud by underwriters, accountants, brokers, the bond counsel, and WPPSS itself. In a prepared statement issued on November 30, 1983, SEC General Counsel, Daniel Goelzer said: "The SEC staff is investigating whether there have been violations of the Federal securities laws in connection with transactions in WPPSS bonds." According to SEC's enforcement chief, John Fedders, the SEC was investigating whether WPPSS concealed or misstated any significant facts about its finances when it issued the bonds, "for instance, when did [WPPSS] know about cost overruns when building the plants, and whether information about this was promptly disclosed to bondholders."

On February 22, 1984, the SEC charged the New York City securities firm of Multi-Vest Securities, Inc., and its two top officers with operating a "classic boiler room" to defraud purchasers of WPPSS bonds. The case against Multi-Vest Securities was the first enforcement action by SEC to come out of its November 1983 investigation. From September 1982 to November 1983, the SEC charged the defendants and their salespeople made false statements or omitted pertinent facts in selling WPPSS bonds. According to John H. Sturc, deputy chief litigation counsel for the SEC, Multi-Vest Securities were accused of misleading customers on the safety and suitability of investing in WPPSS securities, WPPSS' precarious financial condition, the likelihood of a default, the prospect of government assistance in case of a default, the investment ratings and market prices of certain WPPSS bonds and legal disputes over the nuclear projects. In its complaint, the SEC said that Multi-Vest Securities encouraged their 25-person sales staff to use "high-pressure sales techniques" in making "unsolicited telephone calls to prospective customers." Many of the salespeople were ill-trained, inexperienced and also unregistered with the National Association of Securities Dealers, the SEC alleged. SEC investigators detected at least 280 transactions during the first three months of 1983 in which Multi-Vest Securities sold WPPSS bonds at prices 10% to 39.4% higher than the prices it paid for them. The SEC regards any markup above 10% as fraudulent.

The possibility cannot be taken lightly that this litigation will result in judgments against the 87 Participants, the members of WPPSS, and others for the full amount of the $2.25 billion in bonds issued to finance WNP 4 and 5, and possibly for an even larger sum. Such judgments would likely be joint and several, resulting in enormous potential burden on utilities having the ability to pay. They might also have to pick up the burden of the other Participants who have no ability to pay.

The Chemical Bank also asserts that, to the extent that WPPSS and the 88 Participants do not pay the amount of the WNP 4 and 5 bonds, it will seek payment from the Participants in WNP 1, 2, and 3 and from any of their assets. This would include the City of Seattle. Similar assertions can be expected from the plaintiffs in the various class actions filed on behalf of the purchasers of WNP 4 and 5 bonds.

Additional suits are certain to be filed. Defending the litigation will cost the utilities and ratepayers of the Pacific Northwest millions of dollars each year. During the long period in which these cases are pending, the possibility, and possible magnitude, of adverse judgments will probably have to be disclosed in these utilities' financial statements.

Most of the pending cases involve, in one way or another, attempts by WNP 4 and 5 bondholders to recover their losses, or by 4 and 5 Participants to pass to other parties (for example, BPA) any liability they might have to pay the bonds. However, some cases involve other issues. For example, a case on appeal from the U.S. District Court for the District of Oregon tests the validity of the net-billing contracts that are part of the financial underpinning of WNP 1, 2, and 3. The argument against validity is that if, as held by Washington, Oregon, and Idaho courts, the Participants in WNP 4 and 5 had no authority to enter into "hell or high water" contracts with WPPSS on WNP 4 and 5, then the Participants in WNP 1, 2, and 3 had no authority to enter into net-billing contracts. Thus, it is argued, the Participants who signed their "hell or high water" contracts to BPA as an essential part of their net-billing agreements with BPA had only an invalid contract to assign, and BPA acquired no rights or obligations with regard to WNP 1, 2, and 3.

The four investor-owned companies, who own 30 percent of WNP 3, are suing WPPSS and BPA, alleging that they are prepared to finance the completion of their 30 percent share of WNP 3 and that BPA has the legal duty, through net-billing or direct use of its current revenues, either to finance completion of the other 70 percent share in the plant or to repay to the investor-owned utilities the $737 million they have invested in the plant.

The Participants in WNP 4 and 5 and the Chemical Bank are seeking to recover from WPPSS, BPA, and the four investor-owned companies involved in WNP 4 sums as great as $400 million as a reallocation of shared costs of the "twinned" WNP 1 and 4 and WNP 3 and 5.

Some Participants in WNP 4 and 5 have sued WPPSS to recover "bridge" and "termination" loans of some $68 million that they made in connection with 1981 and 1982 efforts to preserve WNP 4 and 5. There are also various suits initiated by, or against, WPPSS involving claims arising from construction or procurement contracts related to WNP 4 and 5.

For convenience of administration, the principal cases arising out of or related to the WPPSS bond default have been consolidated in whole or in part in the U.S. District Court for the Western District of Washington before Judge Richard M. Bilby. Nevertheless, the prospect is for many years of costly and bitter litigation involving most of the utilities in the Pacific Northwest, dozens of law firms (some as advocates, others as defendants), consulting engineers, bond underwriters, hundreds of individuals who have served as officers and directors of these public utility districts over seven years, and representatives of thousands of bondholders seeking to recover or protect the investment they made in WPPSS bonds.

Indicative of the complexity of the litigation spawned by WPPSS' difficulties is the fact that the U.S. Justice Department established in late 1983 a special office in Portland, Oregon, for the purpose of defending the suit against BPA. Staffing the office initially were four lawyers and ten paralegals. To assist them in trial preparation, some 1.1 million BPA documents from government warehouses throughout the Pacific Northwest have been microfilmed and put into a computer; it is estimated that, in addition, 4 to 5 million WPPSS documents will be assembled and examined. Lawyers joining the office were told to expect to stay in Portland at least four years.

And Just What Are the Bondholders to Do?

When WPPSS went into default on WNP 4 and 5, it was up to Chemical Bank, the trustee for the bondholders, to try to find someone to make good the debt. But as Chemical Bank found, there was no obvious source of funds to back the bonds. Until a solution was found, an estimated 75,000 to 100,000 investors who purchased them, usually in $5,000 denominations, from 1976 until the spring of 1981, face the prospect of holding the bonds without further interest payments or selling them on the secondary market to speculators. (No one knows for sure how many investors there are, because all but 4 percent of the $2.25 billion in bonds were issued "To Bearer.")

Very Little Spillover and Overwhelming Silence

For being the largest default in municipal bond history, WPPSS' default on WNP 4 and 5 bonds hasn't had much of a damping effect on the bond market. If anyone was expecting a shock wave on Wall Street, they were wrong. Nor is there likely to be one. The market has taken the default pretty much in stride. One reason the market has shown so little negative reaction, analysts say, is investors' continuing appetite for tax-free bonds. Another reason is that major players in the bond market had long expected and were ready for the default. A third reason for the lack of a strong market reaction is the fact that only a small percentage of WNP 4 and 5 bonds were purchased by banks, insurance companies, and institutions. Had these large, professional investors been more into

WPPSS WNP 4 and 5 it is likely additional efforts would have been made to prevent the default or to recover their investment afterward. For example, concerned that WPPSS' devalued bonds could have a major impact on bank earnings, the comptroller of the currency in Washington, D.C., in the fall of 1983 conducted a major study of 75 percent of the nation's federally chartered banks. The study concluded that none had significant holdings in WNP 4 and 5 bonds; in fact bank holdings were "remarkably slim, with no more than 10 percent of the capital funds of any bank except one," according to Owen Carney, the comptroller's director of investment securities. The study also found that the total holding of the banks surveyed that WNP 4 and 5 bonds amounted to a face value of about $80 million but was worth only a market value of about $12 million in late 1983. It was thought that even relatively small holdings of WPPSS bonds could have an impact on earnings, so Federal regulators ordered the banks that held WNP 4 and 5 bonds to immediately charge off the difference between market and book value against their fourth quarter (1983) earnings.

For the moment, at least, it appears that the biggest losers in the WPPSS fiasco have been the individual bondholders. All this could change, of course, and fairly quickly, if WPPSS' continuing problems threaten the more than $6 billion of bonds issued to finance WNP 1, 2, and 3. This could happen through court judgments against WPPSS. As noted, numerous lawsuits have been filed against WPPSS, its members, participants, financial advisors, and consulting engineers, as well as bond underwriters and bond counsel. A class action suit that consolidated 20 separate bondholders' suits alleges fraud, misrepresentation, and negligence. Bond trustee Chemical Bank has filed a similar suit against WPPSS and member utilities. Both cases are to be heard by the same Federal court, in the summer of 1985, according to the judge. If successful, the lawsuits would have major repercussions for the bond industry, affecting anything from underwriting procedures to disclosure requirements. Although WPPSS has consistently said it would not consider bankruptcy, and although it cannot be forced to take such action, voluntary bankruptcy would provide WPPSS with protection from the ever-growing number of lawsuits. But it could also blur the distinction between bondholders, allowing WNP 4 and 5 holders to try to attach funds intended for WNP 1, 2, and 3. Consequently, many believe that progress in the courtroom offers the best hope of settlement. "There has to be something happening in the courtroom which will prompt people to get together, sit down and talk," says William Berls, a senior trust officer at Chemical Bank.

Nuclear Plant 4 and 5 Bondholders Group Formed

While the lawsuits against WPPSS accumulate, others look to a political compromise, saying that the courts are too slow and a settlement will take too long. One idea has been the formation of bondholder groups to conduct letter-writing campaigns and political lobbying. The goal is to apply political pressure for a settlement at full value. These groups oppose settlement plans that call for reimbursing the bondholder the difference between the face value and the current market value of the bond—basically the solution called for by the so-called Luce Committee. They want full value for their bonds, including interest from the time the bond was purchased. Chemical Bank and bondholders alike believe that if there is to be a full value settlement, Congress will have to enter the WPPSS controversy. But Congress has been reluctant to get involved—and there has been surprisingly little pressure from bondholders for congressional involvement. (Chemical Bank itself cannot lobby Congress, but it can aid bondholder groups, which can.)

Political lobbying efforts have been stymied by the fact that bondholder's groups (or WPPSS, or that matter) don't know who the bondholders are. As noted earlier, only 4 percent of the $2.25 billion worth of bonds issued were registered; all the rest were issued "To Bearer." (Issuing bonds "To Bearer" is not an uncommon practice. It offers several advantages to bond buyers and issuers alike. For example when the number of bondholders are large, the bonds long-term, and they are expected to change hands several times over the life of the bonds, issuing them "To Bearer" is a good way to cut down on expenses. In addition, issuing bonds "To Bearer" affords buyers flexibility, allowing them to sell or pass the bonds on to anyone, anywhere in the world, exchange controls permitting, without having to go to the trouble of notifying anyone of the change. It also affords buyers a great deal of privacy, since their names do not appear on a list of buyers. For issuers, the practice of not registering bonds means that record-keeping can be kept to a minimum and that the onus of dividend or interest distribution is on the investor. Of course, the bond document itself gives the buyer exact instructions for obtaining interest and eventual payment of principal.) Although individual stockbrokers in each of the 50 states might be able to compile a list of WNP 4 and 5 bond buyers, they don't usually disclose such information. Therefore, the success of lobbying efforts were thought to depend on bondholders coming forth on their own.

Chemical Bank officials and others initially believed that despite all the media attention WPPSS and the default had received in 1983, many investors remain unaware or uninformed about the default. It was also assumed that the many bondholders who had had their WPPSS bonds held in unit trusts would come out of the "woodwork" after January 1, 1984, when the semiannual interest payment, worth about $94 million on WNP 4 and 5 bonds, was not paid to the bondholders. For most WPPSS investors, that would have been the first time they didn't receive interest, and thus their first direct financial loss since WPPSS' default in July 1983. (For an investor holding bonds with a face value of $10,000, bought at 12.5 percent interest and due in the year 2010, that will mean a loss of 625 tax-exempt dollars. It is estimated that the average WNP 4 and 5 bondholder owns $20,000 to $30,000 worth of WPPSS bonds. Chemical Bank's hope that the missed January interest payment would help the bondholder lobbying effort fizzled. It appeared that most of WNP 4 and 5's wealthy bondholders were "writing off" their loss for income tax purposes, moving on to other investment opportunities, and turning their back on the poorer bondholder's efforts to win a settlement.

Nevertheless, one small bondholder group that has formed with the help of Chemical Bank is known as the National WPPSS WNP 4 and 5 Bondholders Committee. Organized from a list of volunteers gathered by Chemical Bank, the group's plans to carry out a large-scale letter-writing campaign and intense lobbying efforts failed to materialize in 1984. "While members of Congress may have heard very little from bondholders in 1983," boasted acting chairman Jonathan S. Krasney, "they sure as hell will now." (Krasney would later resign as committee chairman. As of this writing Chemical Bank's National Bondholders Committee is virtually inactive.) On October 4, 1983, Chemical Bank held a mass meeting for WNP 4 and 5 bondholders. A special satellite-TV hookup allowed an estimated 2,000 bondholders around the country to watch the bank's presentation on the default and learn about efforts to get their money back. Chemical Bank hoped the meeting would help unite the bondholders as a force to demand repayment either through a Federal bailout or higher electric rates in the Pacific Northwest. To that end, the bank spent about

$200,000 of the relatively small amount of WPPSS money left in two accounts ($72.8 million remaining in an emergency reserve fund, and $25.7 million that WPPSS turned over to Chemical Bank at the time of the default). Chemical held three simultaneous meetings, one at the 4,643-seat Felt Forum (Madison Square Garden) in New York City, a second at the 6,000-seat Seattle Center Arena, and the third at the Hyatt Regency O'Hare in the Chicago area, which had facilities for 2,000. (Only 1,200 people showed up at the Felt Forum, 250 at the Hyatt Regency O'Hare, and 400 on the plywood-covered ice hockey rink of the Seattle Center Arena.) In addition, the session was also picked up on closed circuit television by brokerage houses in Florida, California, and Texas. WPPSS officials in Richland, Washington, also had a satellite receiving station set up so they, too, could witness the exchange between bondholders and Chemical Bank.

At the Felt Forum, soothing music greeted early arrivals, and guards checked bags for cameras and/or tape recorders, which were not allowed inside. To the surprise of those attending in New York City, where Chemical Bank has its headquarters, those presenting the closed-circuit television program did not speak in person, but instead broadcast live from a midtown Manhattan studio. Grim-faced Chemical Bank officials began the two-hour, twenty-minute session by recounting the events leading up to WPPSS' default the previous July. Although they did take questions from the audience at any location, the questions had to be presented in writing, and were telephoned to the Manhattan studio by offstage bank personnel. This gave bondholders no opportunity to vent their anger, let alone to ask follow-up questions to vague answers.

The five Chemical Bank officials on the program did not offer much hope to the bondholders. They reported that legal fees between January 1981 and July 1983 totaled $3.2 million and were continuing to mount. They said that the bank currently held only three cents for every dollar owed bondholders. They also estimated that perhaps $370 million worth of assets from the two unfinished nuclear units could be sold for only ten cents on the dollar. Chemical Bank pinned most of its hopes for a full settlement on an organized effort by bondholders, presently scattered around the country. Officials said they would help organize a bondholders' committee to make recommendations and advise the bank on possible courses of action or settlements. Chemical Bank Vice President for Investment Securities William Berls urged bondholders to write to their elected representatives. "This is not the Wall Street community's problem," he said. "It is of national import." According to Brian McGirl, another Chemical Bank vice president, "What we are looking for is that the voices of the bondholders are heard, and heard in an intelligent manner." "Clearly, it isn't the role of the trustee to agitate in the political arena, but it could be the role of the advisory committee."

One of the most frequently asked questions was why Chemical Bank had not sued firms that underwrote the bonds—without proper investigation, some contended. "The focus of Chemical's action has been against those primarily responsible for putting the projects together and those who promised to pay the bonds when due," answered Berls. Chemical Bank attorney William Mines, also on the program, noted that a separate class action suit filed on behalf of bondholders had named the four Wall Street underwriters that had marketed the bonds and that this separate action was being consolidated with the Chemical Bank suit.

By the end of the televised session, Chemical Bank had collected the names of more than 100 people interested in serving, without pay, on a bondholders advisory committee. Kenneth L. Dowd, senior vice president of Chemical, said that he expected the committee to become "a powerful voice that will be heard in all the forums" considering possible settlement solutions. To keep bondholders informed, the bank established a special toll-free telephone line for the latest news, updated semiweekly. For the names, addresses and telephone numbers of Chemical Bank officials, advisory committee members, and the toll-free telephone line see Appendix A.

In addition to the official Chemical Bank advisory committee, several other WNP 4 and 5 bondholder committees have been established. One of the first to be organized anywhere was one by Port Angeles, Washington, bondholder William "Bill" Newgent. Committee organizers are not quite sure what it will take to pay the bondholders off in full. But they are united in their efforts to see that "the welchers" are punished. Their main avenue of effort is "to influence public opinion and that of the politicians and government officials with the power to make good on the WPPSS WNP 4 and 5 bonded debt." To that end, Newgent started a monthly newsletter containing information that he and others would need to get their investment money back. "The welchers should be punished," Newgent writes again and again. "This country was built on paying your debts and honoring contracts you signed. A contract's a contract, whether it's an automobile, a house, or WPPSS. The people who loaned the money loaned it in good faith." (As of this writing, Bill Newgent has ceased publication of his newsletter blaming its demise on the lack of sufficient bondholder support, an inactive National Bondholders Committee, and disinterest from Chemical Bank.)

The Bonds: Hold, Sell, or Buy?

If most WPPSS bondholders had known then what they know now, they say, they wouldn't have invested in any of the five nuclear projects. Those who still hold bonds find that they are worth about 15¢ for every dollar they initially paid. These bondholders are worried and angry. Their investment makes headlines, and the news has not been good. Since WPPSS' default on WNP 4 and 5 in July 1983, Chemical Bank has received about 100 telephone calls each business day, and sometimes over 400 calls on days when a significant development occurs. The callers express confusion and anger. They ask what they should do about their WPPSS investments – advice that Chemical won't give. Instead the bank promises only that it will "see the WPPSS fiasco through," with whatever litigation necessary. They remind the callers about the bank's recently formed "bondholders advisory committee" and its efforts to unite bondholders in order to bring political pressure for a speedy settlement. But the bank doesn't give out a great deal of optimism. The court fights will take a long time, officials say, and so will a political settlement. By and large, investment counselors advise the bondholders to their WNP 4 and 5 bonds, for a while longer at least. Since their investments have lost most of their value, why sell them now at 15¢ or 12¢ on the dollar? Who knows, perhaps in six months, maybe 12 to 18 months, a settlement will be reached and they will get a larger return on their investment.

Small Investors' Lessons Learned

Robert Lamb, a professor at the New York University Graduate School of Business Administration, says "statistically, the municipal bond market is riskier today than it has been since the 1930s." He says the risks include the possibility that bonds will suddenly drop in value because of an interest-rate move or a rating change. The risk of default has increased. And bondholders may face the prospect of taxpayer revolts or court challenges over legal technicalities. Also, the July 1983 WPPSS default has to raise serious questions about the way in which municipal bonds are issued, rated, and sold, says Lynn Asinof of *The Wall Street Journal*. According to Asinof, the once super-safe market has grown larger, riskier, and harder to research; one result of this is the growing popularity of insured bonds, on which payment of interest and principal is guaranteed. At the same time, banks and insurance companies, once the major buyers, have been largely replaced by individual investors who are novices in the tax-exempt market.

Another problem is the proliferation of complicated bond issues and credit supports that obscure who is ultimately responsible for repaying the debt. In the past seven years, the municipals market has more than doubled as revenue bonds have mushroomed. In 1982, $78.3 billion of new municipals came to market, up from only $35.4 billion in 1976, the last year that general-obligation bonds predominated. Nearly three-quarters of 1982 new issues were revenue bonds. And the revenue-bond era, says Asinof, has brought a bumper crop of new agencies created to issue bonds. Often, several bodies join to tap the municipals market, further blurring the lines leading back to the ultimate source of the revenue. The result, says Asinof, is a market so complex that even experts admit they often don't understand all the issues they are trading.

It appears, too, that many players in the municipals market are not very knowledgeable. Asinof writes that with inflation pushing more middle-income families into higher tax brackets, growing numbers of individuals are buying municipals to avoid taxes. At the same time, the traditional buyers of municipals—property and casualty insurers and banks—have moved to the side-lines. Banks are less interested because of changed Federal tax laws and have found other ways to shelter profits. As a result, says Robert Lamb, 82 percent of the tax-exempt bonds are sold to individuals, many of whom do not have any idea what they are buying. They buy on a telephone call, he says. They don't read all the documents. Some investors themselves admit that they don't know enough about the bonds. Faced with some 50,000 bond issuers and an estimated 1.4 million issues of varying maturities and yields, they say they have no good way to determine whether a bond is a good buy—even if they look up the ratings on the issue.

Moreover, says Asinof, many bonds are sold to individuals by stockbrokers, who don't have much incentive to invest the time needed to research municipal bonds properly. Says Gedale

Horowitz, a managing director at Salomon Brothers, Inc.: "In the stock market, you trade. In munis [municipal bonds], you don't. It's a one-time commission, and that is it."

It appears that the municipal bond market needs more Federal regulation. Currently it is far less regulated than the stock market. The primary rule-maker in the municipal bond market is the industry's own Municipal Securities Rulemaking Board, which is financed by assessments on dealers. So far, the municipal bond market has beaten back most efforts to impose Federal regulation, although underwriters and issuers are subject to Federal antifraud laws. Asinof believes the municipal bond industry is not likely to emerge from the WPPSS debacle without some regulation. She says that the market's sheer size and complexity warrants at least a strengthening of the rulemaking board. Others in the industry see the charges of fraud and insufficient disclosure surrounding WPPSS as an open invitation for Congress to intervene. "I think the market will, and probably should, be more regulated than it has been," says Goldman Sachs & Company partner Frank P. Smeal. "It was easier when we were selling bonds to banks. You don't have to worry about banks, but you do have to worry about individual investors."

And WPPSS – Who Is to Blame?

Since termination of WNP 4 and 5 in January 1983 and the default on $2.25 billion worth of bonds in July of the same year, there has been no shortage of people – groups, politicians, ratepayers, investors, and others – looking for someplace to pin the blame for the fiasco. Certainly there is no shortage of scapegoats. Steve Lachowicz editorialized entertainingly in *The Wenatchee [Washington] World* on who should be blamed for the mess. I have used some of his accusations combined with my own observations. As for assigning blame for WPPSS it now can be said:

Blame the members of the Washington Public Power Supply System board of directors, those part-time representatives appointed by the local utilities who participated in the projects, and who naively tackled such a monstrous job and then failed to work hard enough to understand what was happening when trouble first surfaced. For them it was much easier to be just a rubber stamp.

Blame WPPSS project management which failed to control

contractors, failed to demand performance from workers, failed to keep their board members and the public accurately abreast of the construction situation and spent frivolously, worrying more about how bond buyers on Wall Street were reacting than how ratepayers in the individual utilities would ever pay the bills.

Blame the attorneys and financial planners who concocted WPPSS' "fair weather" capitalization of interest nuclear plant financing scheme. Good only if a project is completed on schedule and within budget, which most nuclear plants have not been, this "Ponzi" type con-game put everyone in a life or death struggle to complete the projects on time, a huge gamble on their part and generally unknown to ratepayers and bondholders alike. As a result, prior to terminating WNP 4 and 5, of every dollar WPPSS borrowed, sixty cents went to pay off previous interest.

Blame the Washington State public bidding system designed to guard against corruption and making costs of nuclear construction unnecessarily high. How much cheaper might the plants have been if the same design could have been used in all five units, if managers had been free to put contractors they knew to be competent in such work on the job instead of being forced to take the "cheapest" according to bids.

Blame greedy contractors for taking advantage of the cost-plus arrangements and for selfishly capitalizing on what seemed to them to be an endless money supply.

Blame labor for petty jurisdictional disputes which caused work stoppages costing millions of dollars. Also *blame* them for lazy performances bred by the feeling that there was an endless supply of money bankrolling the projects

Blame the United States Nuclear Regulatory Commission for not doing its job with other nuclear plants throughout the country and for sending construction costs skyrocketing with countless change orders and safety regulations. *Blame* the NRC for failing to require adequate training for nuclear plant operators and failure to inspect original nuclear plant designs critically enough so that the accident at Three Mile Island happened. And *blame* that accident for unnecessarily putting a few more nails in the coffin of nuclear power in the United States.

Blame the public for failing to be informed enough about nuclear power and for being lazy about following the actions of their public utility district's boards of directors when the WPPSS projects were being planned, the participant agreements signed, and what might be involved in financing all five units at the same time through a "Ponzi" like scheme called capitalizing the interest.

Blame the broadcast and print media for being largely unconcerned about the fact that only large organizations with personnel experienced in the construction and operation of nuclear plants should be allowed to tackle such a large, complex project as to build and operate even one single nuclear plant, let alone a long-term, multiplant construction operation.

Blame the Washington State legislature for giving WPPSS "blank check legislation" to construct energy projects without having to answer to anyone, without the customary checks and balances, and without allowing themselves an oversight role over WPPSS' activities.

Blame elected officials at the state and Federal levels for perpetuating spending patterns and budget deficits which sent the economy into a nose dive in the 1970s and left inflation spiraling up at such rates as to make revised construction cost estimates unthinkable and the nuclear plants no longer affordable in the eyes of most people.

Blame the energy planners in Bonneville Power Administration, the investor-owned utilities, the Washington state energy agencies, the public utility districts, etc., for believing that the consumption of electricity would grow forever, that electric energy planners know everything and the general public like those in Seattle know nothing.

Blame the investment community for failing to verify the legal validity of "hell or high water" contracts as a prior condition for the advance of $2.25 billion in individual investor monies and then telling them that their total risk was minimal without bothering to do proper research. Contracts which are too onerous are contracts made to be broken. *Blame* the investment community also for accepting enthusiastic opinions from their own research divisions and from bond rating services without being probing and skeptical of a power entity with no track record in the power business and setting out to build no fewer than five nuclear plants at the same time.

Blame the tens of thousands of bondholders, most of whom didn't have the slightest idea of what WPPSS was and its tax-exempt bonds they were buying. *Blame* them for being trusting in purchasing WPPSS bonds only on the basis of a telephone call from their broker instead of reading all the documents they should have to determine if WPPSS and/or its bonds were any good or not prior to investing their money.

Blame the bond rating agencies for a failing rating system which gave Washington Nuclear Plants 4 and 5 bonds a continuous healthy rating of A1 and A+ as late as March 1981, seven years after WPPSS had begun its pattern of massive cost overruns. When the bond rating services say a bond is risky, believe them. When they say a bond is not risky, don't bank on it.

Blame the Federal government for watching the proliferation of municipal revenue bonds in recent years and allowing the municipal bond industry to go virtually unregulated.

Blame the refusal of originally enthusiastic participants in WPPSS to pay their signed contract share of continuing costs. While the denial of any financial responsibility by WPPSS' public power Participants may be legal, it appears that even signed contracts don't mean anything should the going get tough or expensive.

Blame Congress for its continuing "hands off" attitude on averting the biggest municipal bond default in history. And *blame* them for not taking the time to work out a post default settlement, leaving the Pacific Northwest struggling for years under a massive blanket of lawsuits.

Blame the American lifestyle which demands all that electrical energy in the first place to "improve" the level of comfort at home, work, and recreation, an attitude which too readily tolerates waste in all forms, not just in energy.

Blame everyone and therefore no one.

Background Notes

Introduction

New York Times, "Shoreham Epitomizes End of Nuclear Building Era," Dec. 8, 1983.
The New York Times, "Utilities with Nuclear Woes," October 14, 1983.
The Wall Street Journal, "Utilities Face a Crisis Over Nuclear Plants; Costs, Delays Mount," page 1 (East), November 28, 1983.
Time Magazine, "Nuclear Power Bombing Out?" (Cover Story), Feb. 13, 1984, p. 34.

Chapter 1. The Setting

Atomic Energy Act of 1946, U.S. Statutes at Large, v.60 (1946).
Atomic Energy Commission, Appropriation, U.S. Statutes at Large, v.76 (1962).
Arthur Holly Compton, *Atomic Quest*, Oxford University Press, New York, 1956.
Vera Edinger, *Hanford: A Success Story Against Tremendous Odds*, Washington Public Utility Districts Association, Seattle, 1964.
Richard G. Hewlett and Francis Duncan, *Atomic Shield 1947–1952: A History of the U.S. Atomic Energy Commission*, Volume 1 and 2, The Pennsylvania University Press, University Park, PA, 1969.
Darryll Olsen, "The Washington Public Power Supply System: The Story So Far," (Research Paper), Office of Applied Energy Studies, Washington State University, Pullman, 1981.
Tri-City Herald, Richland, WA, September 27, 1963.
U.S. Bonneville Power Administration, *Columbia River Power for the People: A History of Policies of the Bonneville Power Administration*, U.S. Government Printing Office, Washington, D.C., 1981.
U.S. Bonneville Power Administration, *The Hanford Project: A Legislative History*, Portland, OR, 1967.
U.S. Energy Research and Development Administration, *Hanford: Yesterday, Today, and Tomorrow*, Richland, WA, 1975.
Washington (State), *Revised Code of Washington*, Chapter 43.52.

Chapter 2. Need for Power

Darryll Olsen, "The Washington Public Power Supply System: The Story So Far."
Theodore Berry & Associates, *Management Study of the Roles and Relationships of Bonneville Power Administration and Washington Public Power Supply System*, Seattle, 1981.
U.S. Bonneville Power Administration, *A Ten-Year Hydro-Thermal Power Program for the Pacific Northwest*, Portland, OR, 1969.
U.S. Bonneville Power Administration, *Columbia River Power for the People: A History of Policies of the Bonneville Power Administration*.

Washington State University, *Final Report to the Washington State Legislature*, "Independent Review of Washington Public Power Supply System Nuclear Plants 4 and 5," Office of Applied Energy Studies, Washington Energy Research Center, Pullman, WA, 1982.

Chapter 3. Bonneville Power Administration and the Washington State Public Utility Districts

Charles F. Luce, Edward E. Carlson, and Herbert M. Schwab, "A Report to Governor John Spellman of Washington and Governor Victor Atiyeh of Oregon: Recommendations for Solutions to the Major Problems Involving, or Arising from the Washington Public Power Supply System," Portland, OR, November 22, 1983.

Darryll Olsen, "The Washington Public Power Supply System: The Story So Far."

U.S. Bonneville Power Administration, *A Ten-Year Hydro-Thermal Power Program for the Pacific Northwest*.

U.S. Congress, House, Subcommittee on Mining, Forest Management, and Bonneville Power Administration, Committee on Interior and Insular Affairs, Representative James Weaver (D–Oregon) presiding "Oversight Hearing on Bonneville Power Administration," Washington D.C., February 14–15, 1983.

U.S. General Accounting Office, *Bonneville Power Administration and Rural Electrification Administration Actions and Activities Affecting Utility Participation in Washington Public Power Supply System Plants 4 and* 5, Washington, D.C., July 30, 1982.

Chapter 4. Seattle Residents Order Their City Light Company Not to Participate in WNP 4 and 5

Peter Henault, Washington, D.C., Interview with author, January 1978.

Walter L. Williams, Assistant Corporation Counsel, The City of Seattle, Seattle, WA, Interview with author, February 1978.

Joel Connelly, "Seattle's 'No' to WPPSS Looks Wise Now," *Post-Intelligencer*, Seattle, WA, Page B-4, January 17, 1982.

Joel Connelly, "2 Seattle Officials Favor WPPSS Aid," *Post-Intelligencer*, Seattle, WA, January 26, 1982.

Chapter 5. The WPPSS' Management Organization and Philosophy

Darryll Olsen, "The Washington Public Power Supply System: The Story So Far."

Theodore Barry & Associates, "Management Study of the Roles and Relationships of Bonneville Power Administration and Washington Public Power Supply System."

Charles F. Luce, etc., *A Report to Governor John Spellman of Washington and Governor Victor Atiyeh of Oregon*.

John Hayes, "Falling Power Demand, Ballooning Costs Stun Region," *The Oregonian*, Portland, OR, Page C2, September 18, 1981.

U.S. General Accounting Office, "Bonneville Power Administration and Rural Electrification Administration Actions and Activities Affecting Utility Participation in Washington Public Power Supply System Plants 4 and 5."

Washington State Senate Energy and Utilities Commission, *Causes of Cost Overruns and Schedule Delays on the Five WPPSS Nuclear Power Plants*, Vol. I (Report to the Washington State Senate and the 47th Legislature, Olympia, WA, January 1981).

Chapter 6. The BPA Role and Relationship with WPPSS

Theodore Barry & Associates, "Management Study of the Roles and Relationships of Bonneville Power Administration and Washington Public Power Supply System."

John Hayes, "WPPSS Haunted by Closed Doors," *The Oregonian*, Portland, OR, September 14, 1981.

John Hayes, "Patriarch Resents Outside Interference," *The Oregonian*, Portland, OR, September 14, 1981.

Paul Manley, "NW Power Forecasts 'Discredited,'" *Oregon Journal*, Portland, OR, November 7, 1981.

Chapter 7. The WPPSS "Capitalization of Interest" Financing Scheme

Bruce Ingersoll, "New York Broker Charged by SEC on WPPSS Sales," *The Wall Street Journal*, Page 3 (East), February 22, 1984.

Washington State University, *Final Report to the Washington State Legislature*.

Darryll Olsen, "The Washington Public Power Supply System: The Story So Far."

Donald J. Sorensen, "Fate of 2 WPPSS Nuclear Plants Hinges on Financing," *The Oregonian*, Portland, OR, Page 2M, September 15, 1981.

Donald J. Sorenson, "Utilities Discontented as Cost Woes Escalate," *The Oregonian*, Portland, OR, Page C3, September 18, 1981.

John Hayes, "Falling Power Demand, Ballooning Costs Stun Region."

Richard L. Hudson, "SEC Is Probing Range of Possible Culprits in Default on $2.25 Billion WPPSS Bonds," *The Wall Street Journal*, November 30, 1983.

Chapter 8. The WPPSS Is Accused of Mismanagement

Washington State Senate Energy and Utilities Committee, *Causes of Cost Overruns*.

John Hayes, "Regulatory Changes Cause Ripple Effect Forcing Untold Delays," *The Oregonian*, Page 19, September 18, 1981.

John Hayes, "Organization Woes Compounded Plant Construction Troubles."

Duff Wilson, "Old Woes, New Leaders, Bigger Outlay for WPPSS," *The Herald*, Everett, WA, Page 6A, May 9, 1981.

Bob Lane, "Inflation: Equipment Small Part of N-plant Costs," *Times*, Seattle, Washington, July 4, 1981.

Bill Stewart, "WPPSS Fudged Estimate on Plant," *The Columbian*, Vancouver, WA, September 22, 1981.

Bill Curry, "Costs of Nuclear Plants Dry up Profits," *The Bulletin LA Times*, Los Angeles, CA, no page, no date.

Darryll Olsen, "The Washington Public Power Supply System: The Story So Far."

"WPPSS Balks on Data," *The Daily News*, Port Angeles, WA, Page A6, March 11, 1982.

Jack Briggs, "WPPSS Settles with Contractor It Evicted," *Tri-City Herald*, Richland, WA, Page 2, February 4, 1982.

Joel Connelly, "Cost of WPPSS Contract Rockets by $300 Million," *Seattle Post-Intelligencer*, December 2, 1981.

Peyton Whitely, "Long Face: Insulator's WPPSS Contract Went Sour," *The Seattle Times*, Page C4, October 18, 1981.

Joel Connelly, "WPPSS Increases Satsop Contract by $100 Million," *Seattle Post-Intelligencer*, January 30, 1982.

"WPPSS Reportedly Pays Disputed $870,000 Despite Questions," *The Sunday Oregonian*, Page C5, November 1, 1981.

Sandy Nelson, "WPPSS–Ebasco Pact Detailed," *Daily World*, Aberdeen, WA, September 13, 1981.
Joel Connelly and Gil Bailey, "WPPSS Paid Big Expenses After Auditors Said No," *The Seattle Post-Intelligencer*, October 31, 1981.

Chapter 9. The Lure of Tax-Exempt Bonds

"The Yield's the Thing in Tomorrow's Bond Market," *Business Week*, Page 144, December 26, 1983.
Donald J. Sorenson, "Fate of 2 WPPSS Nuclear Plants Hinges on Financing," *The Oregonian*, Portland, OR, Page B2, September 15, 1981.
"Crash of '82? WPPSS Problems Snowballed," *The Seattle Times*, Page C3, January 17, 1982.
Robert Lamb and Stephen P. Rappaport, *Municipal Bonds*, McGraw-Hill, New York, 1980.
James A. Maxwell and J. Richard Aronson, *Financing State & Local Governments*, The Brookings Institution, Washington, D.C., 3rd Edition, 1977.
David M. Darst, *The Complete Bond Book*, McGraw-Hill, New York, 1975.
Lynn Asinof, "Growth of Municipals Market and of Role of Small Investors Raises Complex Issues," *The Wall Street Journal*, Page 60 (East), November 4, 1983.
Washington State University, *Final Report to the Washington State Legislature*.
Charles F. Luce, et al., *A Report to Governor John Spellman of Washington and Governor Victor Atiyeh of Oregon*.
Jane Bryant Quinn, "The Lure of Tax-Free Bonds," *Newsweek*, Page 76, Dec. 19, 1983.
Howard Rudnitsky, "Whoops," *Forbes*, Page 60, June 6, 1983.

Chapter 10. Initiative 394

Joel Connelly and Larry Lange, "N-Curb Passes but Court Fight Looms," *Seattle Post-Intelligencer*, November 4, 1981.
Doug Underwood, "394's Margin Surprised Backers," *The Seattle Times*, Page B2, November 5, 1981.
Jim Dullenty, "Voter-Approved 394 Faces Court Challenge," *Tri-City Herald*, Pasco, WA, November 4, 1981.
"The Case Against 394," *The Everett Herald*, October 18, 1981.
Joel Connelly, "The Man Behind N-Curb Campaign," *Seattle Post-Intelligencer*, November 9, 1981.
Walter Hatch, "Massive Donations from Industry Make Anti-394 War Chest Biggest," *The Daily News*, Port Angeles, WA, October 27, 1981.
John Hayes, "Initiative Gives Ratepayers Veto Over WPPSS Bond Sales," *The Oregonian*, no date.
Joel Connelly and Joe Frisino, "Banks Sue to Overturn Vote," *Seattle Post-Intelligencer*, December 5, 1981.
Victor F. Zonana, "Rebellion Breaks Out in Northwest over Skyrocketing Electricity Rates," *The Wall Street Journal*, March 19, 1982.
Joel Connelly, "That WPPSS Vote May Bother Investors," *Seattle Post-Intelligencer*, November 5, 1981.

Chapter 11. Washington Plants 4 and 5 Terminated

"Dan Evans on the NW's Energy Future," *Northwest Energy News*, Northwest Power Planning Council, Portland, OR, March 1982.

John Gillie, "Bail Out WPPSS? Centralia Votes No on Mothball Plan," *The News Tribune*, Tacoma, WA, September 30, 1981.
Steven Smith, "SUB Shuns New Rate Hike," *Register-Guard*, Eugene, OR, October 20, 1981.
Darryll Olsen, "The Washington Public Power Supply System: The Story So Far."
John Gillie, "Thursday Is D-Day for N-Plants," *The News Tribune*, July 22, 1981.
Peyton Whitely, "WPPSS Budget: All 5 Plants, Despite Cloud," *Seattle Times*, July 25, 1981.
Victor F. Zonana, "Two Washington Public Power Units Would Be Mothballed in Utilities' Plan," *The Wall Street Journal*, September 25, 1981.
Bob Lane, "3 Resignations at WPPSS Won't Affect Management, Board Is Told," *The Seattle Times*, Page A7, January 16, 1982.
Leslie Wayne, "Halt Urged at Nuclear Project," *New York Times*, January 16, 1982.
Steve Boyer, "Local Utilities Back Plan to Mothball WPPSS Plants," *Bulletin*, Bend, OR, December 20, 1981.
Washington State University, *Final Report to the Washington State Legislature*.
Les Blumenthal, "WPPSS Puts Off Completion of Second Plant," *Missoulian*, July 9, 1983.
John Hayes, "Weaver: Vocal Thorn in Supply System Side," *The Oregonian*, Portland, OR, Page B3, September 14, 1981.
Joel Connelly, "Utilities Agree on 2 N-Plants," *Seattle Post-Intelligencer*, October 23, 1981.
"Financier Supports WPPSS Bonds," *Seattle Daily Journal of Commerce*, July 24, 1981.
Robert Caldwell, "Utility Board Approves N-Plant Mothball Plan," *News*, Springfield, OR, September 29, 1981.

Chapter 12. Litigation

Carrie Dolan, "Washington State Justices Say Utilities Not Liable for $2.25 Billion WPPSS Debt," *The Wall Street Journal*, Page 3 (East), June 16, 1983.
"The Fallout from 'Whoops,'" *Business Week*, Page 80, July 11, 1983.
Dan Trigoboff, "WPPSS Ruling: Confusion in Contracts Law," *Sonoma County Herald-Recorder*, Santa Rosa, CA, July 12, 1983.
John P. McGowan, "Municipal Joint Agency Contracts: The Holes in 'Dry-Hole' Provisions," *Public Utilities Fortnightly*, Page 49, October 13, 1983.
"WPPSS Might Levy Property Tax to Pay off Debts," *Register-Guard*, Eugene, OR, May 17, 1983.
"Spellman Says Odds Against WPPSS Bill," *News*, Port Angeles, WA, May 22, 1983.
Washington State University, *Final Report to the Washington State Legislature*.
Charles F. Luce, et al., *A Report to Governor John Spellman of Washington and Governor Victor Atiyeh of Oregon*.

Chapter 13. Chemical Bank vs. Washington Nuclear Plants 4 and 5 Participants

"Chemical Bank Asks Court to Reconsider Decision on WPPSS," *The Wall Street Journal*, July 6, 1983.
Joel Connelly, "Default by WPPSS to Jolt State Projects," *Seattle Post-Intelligencer*, July 26, 1983.
Les Blumenthal, "WPPSS Ruling Evokes Cheers, Dire Warnings," *World*, Wenatchee, WA, Page 1, June 16, 1983.
"Spellman Releases WPPSS Plans," *News*, Port Angeles, WA, May 5, 1983.

Carrie Dolan, "WPPSS Board Is to Meet on Critical Issues of Possible New Loans, Bankruptcy Filing," *The Wall Street Journal*, June 24, 1983.
David Ammons, "Spellman's Move to Head Off Default Is Met with Skepticism by Democrats," *Aberdeen Daily World*, Aberdeen, Washington, May 5, 1983.
Charles F. Luce, et al., *A Report to Governor John Spellman of Washington and Governor Victor Atiyeh of Oregon*.
Washington State University, *Final Report to the Washington State Legislature*.

Chapter 14. Default and Its Aftermath

Charles F. Luce, et al., *A Report to Governor John Spellman of Washington and Governor Victor Atiyeh of Oregon*.
Darryll Olsen, "The Washington Public Power Supply System: The Story So Far."
Lynn Asinof, "Growth of Municipal Market and of Role of Small Investors Raises Complex Issues," *The Wall Street Journal*, Page 60 (East), November 4, 1983.
Carrie Dolan, "Many WPPSS Investors Still Cling to Hope as Default Looms on $2.25 Billion of Bonds, *The Wall Street Journal*, Page 10 (West), May 19, 1983.
Nicholas von Hoffman, "WPPSS Debacle Offers Lesson in Bonds," *News*, Anchorage, Alaska, August 4, 1983.
"Learning from Whoops," *The Wall Street Journal*, Page 4 (East), July 27, 1983.
Steve Lachowicz, "There's No Shortage of Places to Pin the Blame for WPPSS Fiasco," *The Wenatchee World*, Wenatchee, WA, June 2, 1981.
"What Have We Learned from WPPSS Case?" *The News Tribune*, March 21, 1982.
John Hayes, "Falling Power Demand, Ballooning Costs Stun Region," *The Oregonian*, Page C2, September 18, 1981.
Robert A. Lincicome, "The Saddest of the Lessons Learned from the WPPSS Fiasco," *Electric Light & Power Magazine*, Vol. 61, #9, September 1983.
Washington State University, "Final Report to the Washington State Legislature."
William Newgent, Port Angeles, WA, telephone interview with author, January 9, 1983.
"Biggest Participant in 2 WPPSS Units Tests Bond Market," *The Wall Street Journal*, November 15, 1983.
Michael Blumstein, "A Sad Bondholder Session," *The New York Times*, October 5, 1983.
Lynn Asinof, "Bondholders from WPPSS to Meet Today," *The Wall Street Journal*, October 5, 1983.
Lynn Asinof, "WPPSS Begins to Cause Pain for Investors," *The Wall Street Journal*, November 15, 1983.
Susan Gordon, "'Whoops' Bondholder Fights Back," *The Daily News*, Port Angeles, WA, Page 1, October 14, 1983.
Lynn Asinof and David B. Hilder, "WPPSS Bondholders Won't See Money for at Least 3 Years, if at All, Trustee Says," *The Wall Street Journal*, October 6, 1983.
Ralph W. Shaw, General Manager, North Carolina Municipal Power Agencies, Raleigh, NC, telephone interview with author, January 6, 1983.
Joel Connelly, "Senators Look to U.S. Rescue Plan for WPPSS," *Post Intelligencer*, Seattle, WA, Page 1, May 20, 1983.
Joel Connelly, "Senators Act to Rescue N-Plants," *Post Intelligencer*, Seattle, WA, July 19, 1983.
Joel Connelly, "Spellman Gives a Green Light to N-Plant Rescue," *Post Intelligencer*, Seattle, WA, July 18, 1983.
"House Criticizes Senate Financing Plan for WPPSS," *Oregonian*, Portland, OR, July 23, 1983.
"Opponents Claim Nuclear Power Plant Bill Flawed," *Oregonian*, Portland, OR, September 15, 1983.

"Federal Government Should Aid WPPSS Situation: Spellman," *Bee*, Sandpoint, ID, June 22, 1983.

"Senate Leaves WPPSS in Air," *Oregonian*, Portland, OR, August 5, 1983.

Ross Anderson, "McClure Drops Senate Attempt for WPPSS Rescue Funds," *Times*, Seattle, WA, Page 1, September 19, 1983.

Les Blumenthal, "Utilities Seek Congressional Aid to Rescue WPPSS Nuclear Plant," *Missoulian*, Missoula, MT, July 15, 1983.

Mark Funk, "Measure to Aid WPPSS Plant 3 Clears First Hurdle," *Herald*, Everett, WA, July 19, 1983.

Joel Connelly, "N-Plant Bail-Out a Step Nearer," *Post Intelligencer*, Seattle, WA, July 20, 1983.

Sandra McDonough, "Officials Try to Woo Aid for WPPSS," *Oregonian*, Portland, OR, July 14, 1983.

Bibliography

Newspaper Articles

"Connecticut Utility's Backlash," *New York Times*, April 30, 1984.

"Michigan Utility May Abandon Nuclear Project," *New York Times*, April 27, 1984.

"New Plan to Manage Seabook: Partners Act After Pullout of Lead Utility," *New York Times*, April 24, 1984.

"Ohio Utilities Face Nuclear Plant Doubts," *New York Times*, April 30, 1984.

"Louisiana Power Sues Firm for Construction Delays on Nuclear Power Plant," *Wall Street Journal*, December 9, 1983, page 3 (West).

"2 Nuclear Plants to Be Delayed," *New York Times*, November 9, 1983, page 29(N) and page D1(L).

"Nuclear Plant Delay Urged," *New York Times*, October 28, 1983, Page 46(N) and page D14(l).

"Inspecting the Core of Nuclear Power Plants," (letter) Brewer, Shelby T.; Teller, Edward; Weinberg, Alvin M.; O'Loughlin, M.E.J., *Wall Street Journal*, August 17, 1983, page 23 (West) and page 23 (East).

"Pump Damage Delays Arizona Nuclear Plant," *Los Angeles Times*, August 20, 1983, volume 102, section I, page 3.

"The Lousy Economics of Nuclear Power," (column) Caroline J.C. and Hellman, Richard. *Wall Street Journal*, August 2, 1983, page 28 (West) and page 28 (East).

"Zurn (Industries) Says TVA to Pay $15.7 Million for Costs of Construction Delay," *Wall Street Journal*, October 8, 1982, section 2, page 26 (West) and page 56 (East).

"A Utility's Nuclear Emphasis," *New York Times*, August 4, 1982, page 1. Commonwealth Edison Company has long boasted of being a leader in nuclear power generation among the nation's electric utilities. But what it considers to be its strongest point is now causing it considerable hardship because of its ambitious construction program. At one time the utility had hoped to be generating 50 percent of its electric power through nuclear units but its return on equity has dropped to 13 percent from 17.5 percent due to construction delays.

"TVA Will Delay Another Three Nuclear Plants: Utility to Sell $700 Million of Bonds, It's Biggest Issue, to U.S. Agency at 13.565%," *Wall Street Journal*, March 5, 1982, page 2 (West).

"Power Struggles: Environmental Group, in Charge of Strategy, Is Stressing Economics: Defense Fund Says Utilities Can Fill Needs and Save by Foregoing Big Plants: Are Its Estimates Reliable?" *Wall Street Journal*, September 28, 1981, page 1 (West) and page 1 (East).

"Government Regulation May Not Be Main Reason for Power Plants Delays: Study Indicates Other Factors as Equally or More Important in Longer Construction Times," *Christian Science Monitor*, June 23, 1981, page 4.

"A Setback for Nuclear Plants," *New York Times*, June 19, 1981, page 1. Five nuclear power plants under construction in Washington State in the 1970s are still incomplete, caught in a net of rising construction costs, regulatory snarls, and work delays.

"Economics of Nuclear Energy" (letter) Wilson, Richard; Kline, Robert V.; Jaffe, Herbert, *New York Times*, May 31, 1981, section 6, page 94.

"Hard Times for Nuclear Power," *New York Times Magazine*, April 12, 1981, section 6, page 36. Prospects for nuclear power in the U.S. are bleak due to rising costs and repeated delays in the construction of nuclear power units and have seriously damaged the financial structure of many utilities. Declining electricity demand has significantly reduced and in some cases eliminated the need for additional generating capacity.

"Can Reagan Lift the Cloud over Nuclear Power?" *New York Times*, March 8, 1981, section 3, page 1. Regulatory delays, rate restrictions, the unsolved waste disposal problem, lower-than expected demand for electricity, inflation, the loss of public confidence, and the lack of a clear governmental policy have severely impacted the U.S. nuclear industry.

"Continuing Cloud: Midwest Nuclear Plant in the Works 13 Years, Keeps Facing Delays," *Wall Street Journal*, March 4, 1980, page 1. As a result of the Three Mile Island nuclear reactor accident, nuclear facilities throughout the U.S. are in serious jeopardy. New NRC safety requirements are driving up costs and lengthening construction periods of the 90 nuclear plants currently under construction in the U.S. To illustrate the dilemma faced by U.S. nuclear power plants, the situation at Consumers Power Company of Midland, Michigan is described.

"Philippines Reactor Delay," *New York Times*, July 4, 1979, section D, page D2.

"NRC Again Delays Vote on Closing Power Plants," *Wall Street Journal*, April 27, 1979, page 7 (West) and page 7 (East).

Department of Energy (DOE) Database

The database of the United States Department of Energy has one of the world's largest sources of literature references on all aspects of energy and related topics. Their database includes coverage of journal articles, report literature, conference papers, books, patents, dissertations, and translations.

"Nuclear Energy's Future: Lifting the Regulatory Cloud," Walske, C., Atomic Industrial Forum Incorporated, Washington, D.C., September 12, 1983, 18 pages, DOE Accession Number: 1209265 EDB-84:001763. Author states that nuclear energy provides 13 percent of U.S. and 10 percent of world electricity. This has been achieved author says with an exemplary safety record and less insult to the environment than any other power source. Author argues that nuclear power is 15 percent cheaper than coal despite the high capital and regulatory costs, but regulatory delays in the construction and licensing periods have increased 70 percent to 10 to 14 years, more than twice the lead time in France and Japan. Author outlines new legislation for site pre-approval, plant standardization, combined construction and operating licenses, and hybrid procedures for public hearings that would make regulation less uncertain.

"Sizing up Sizewell," *Economist* (London, United Kingdom), April 23, 1983, 38–40 pages, DOE Accession Number: 1189944 AIX-14: 775943, EDB-83:184776. Proposal to construct a nuclear power plant at Sizewell, in Suffolk, is discussed under the headings: paying for power (economics of Sizewell), paying for capital (high costs of nuclear power plants), paying for delays (effects of construction delays and design changes to meet safety requirements), paying for coal (circumstances in which a fossil-fuel power plant would be less costly than expected), paying for time (advantage argueds for postponing Sizewell's construction while pressing on with an energy conservation program).

"Going Broke on Atoms," Loeb, P., March 29, 1982, 19 pages, DOE Accession Number: 981209, EDB-82:156066. The financial mess surrounding four unfinished nuclear plants planned by the Washington Public Power Supply System (WPPSS) with public support has turned the projects into a controversial issue that threatens the future of U.S. nuclear power. Author says the economic effects of construction delays are multiplied by the need to respond to new safety standards and challenges but he also blames WPPSS' young labor force, their billing system which committed ratepayers to cover costs, and poor management for the problems.

"Impact of Licensing Costs in the Federal Republic of Germany," Kraemer, H., October 1981, 34 pages, DOE Accession Number: 902289, AIX-13:662055, EDB-82:077134. Author writes that the gradual cut-backs in oil and natural gas means an increase in the role of nuclear power. Generation costs, he shows, and fuel supply are a powerful argument for nuclear energy. The long term electricity supply forecast for Germany he says is approximately 30,000 MWe of nuclear. At present 9000 MWe is in operation and 10600 MWe is under construction. The costs per installed KW in 1979 is shown to be almost three times as high as in 1971, partly due to licensing delays which cause increased interest charges during construction.

"An Economic Analysis of the Effects of Regulatory Delay on Nuclear Power Plant Construction," Maloney, M.T. and Walsh, M.D., August 1980, 105 pages, DOE Accession Number: 796109 INS-81:015115, EPA-07:005391, EDB-81:104375. Authors believe that in order to evaluate the impact that any government regulation has on society, an accurate measure of the costs imposed by the regulation is essential. The purpose of this report is to define and measure the true impact that construction delays have on total project costs. The nuclear power industry has been chosen to illustrate the problem. A model has been developed to deal with the costs involved in the construction of a typical nuclear power plant in terms of its economic and physical environment. The model is then tested by regression time and cost data of 31 completed plants to determine the impact that unanticipated delays have had on total project costs. These results indicated that such delays would increase the total project costs of a typical 1,000 MW plant by .8 percent per month in the initial stages of the project and 1.1. percent per month after actual construction begins.

"Study of Factors Governing U.S. Utility Nuclear Power Decisions," Stoller (S.M.) Corporation, May 1980, 49 pages, DOE Accession Number: 777286, EPA-07:004348, ERA-81:025482, INS-81:012066. The S.M. Stoller Corporation (SMSC) conducted a study of the U.S. utility attitudes toward nuclear power. In the course of this study SMSC carried out a utility survey the objectives of which were: (1) to identify and rank in importance the governing considerations in actions taken in the past three years to cancel or defer nuclear projects, and (2) to gain insight into the circumstances over the next several years. During the survey, contacts were made at the senior management level with utilities representing approximately half of the nation's total electric capacity and two-thirds of its present nuclear commitment. Analysis of the responses led to the conclusion that most, if not all, of the decisions reached by the respondent utilities in the past several years to cancel or defer nuclear projects were triggered by one or a combination of the following four considerations: financial constraints; reduction in expected system load growth; licensing and construction and/or unpredictability; and adverse state government policies or attitudes regarding nuclear power.

"Nuclear Project Plagued by Cancellations, Delays," March 1980, 3 pages, DOE Accession Number: 626579, EDB-80:066103. Decisions to terminate nuclear power projects and delays in the construction and licensing of those already underway carry some risk that electrical supplies after the mid-1980s may become unreliable.

"Input-Output Model of Regional Environmental and Economic Impacts of Nuclear Power Plants," Johnson, M.H. and Bennett, N.J., George Mason University, Fairfax, VA, May 1979, 252 pages. DOE Accession Number: 582020, INS-80:002058, EPA-06:001121, EDB-80:021542. The costs of delayed licensing of nuclear power plants for a more comprehensive method of quantifying the economic and environmental impacts on a region. A traditional input-output (I-0) analysis approach is extended to assess the effects of changes in output, income, employment, pollution, water consumption, and the costs and revenues of local government disaggregation among 23 industry sectors during the construction and operating phases.

"N-Delays Force VEPCO to Eye Coal Use," Ruth, P., October 22, 1979, 16 pages, DOE Accession Number: 571450, EDB-80:010972. The Virginia Electric Power Company (VEPCO) may convert two practically constructed nuclear power plants to coal because of economic and regulatory uncertainties. The utility may also sell a partially constructed hydroelectric and pumped storage station to raise funds for its other construction projects. A six-month study will examine the cost effectiveness of coal and nuclear units. Some of the construction delays have been due to a slower than expected growth in demand due to conservation and business slowdowns.

"Nuclear Power's Effects on Electric Rate Making," Smith, D.S. and Lancaster, A.A., February 2, 1979, 22 pages, DOE Accession Number: 365765, ERA-03028619, INS-78007491, EPA-04:003100. Government and electric utility industry officials are re-evaluating nuclear power's contribution to the total U.S. energy supplies. This piece addresses how the recent increased nuclear plant construction and operation costs are translated into the prices that consumers pay for electricity. Nuclear power costs are expected to be an even larger portion of the total electricity price due to the type, timing, and magnitude of the nuclear costs involved such as: (1) inclusion of construction work in progress in the rate base; (2) fuel adjustment clauses and treatment of nuclear fuel cycle costs; (3) treatment of certain taxes under the rate-making method called normalization or deferral accounting; and (4) rate treatment for particular nuclear expense items reflecting costs of delays, plant cancellations, and operational slowdowns.

"Nuclear Energy: A Key Role Despite Problems," Anderson, E.W., March 7, 1977, 12 pages, DOE Accession Number: 243986, INS-77:009177, EDB-77:081993. Nuclear energy is projected by the author to be the fastest growing power source and a key to meeting power demands in spite of the many problems facing the nuclear industry in the form of delays, protests, and cancellations.

"Economics of Electricity Generation: The Context of the 1975 California Nuclear Power Plant Initiative," Anderson, S.O., 1976, 110 pages, DOE Accession Number: 243976, ERA-02:035365, EDB-77:081983. Author argues that nuclear power plants are poor investments because of cost escalations, unreliability, component failure, and vulnerability to condemnation for safety reasons. He reviews the distinction between external costs and internal or direct operating costs, and analyzes the internal-cost experience of existing nuclear power plants; he shows that considerable variation exists in such costs, and hence in the reliability of nuclear plants. This lack of reliability, if it persists, he writes, would significantly increase the price of nuclear power, and thus could eliminate any expected cost advantages from building and operating nuclear plants. He believes that funds diverted from nuclear construction would provide more benefits if spent on energy conservation or alternative power supplies.

"Utilities' Experience Relating to the Financing of Nuclear Power Plants," Uhrig, R.E. and Karam, R.A. (editors), 1976, 587 pages, DOE Accession Number: 223628, ERA-02:029971, EDB-77:061463. This paper reviews the experience of utilities in financing nuclear plants and analyzes the major causes of delays and increases in costs. The principal difficulties identified are nuclear regulatory changes, overlapping and duplicative state, Feeral, and regional jurisdictions, lack of standardization in nuclear power plant design, and inflation with the resulting increases in the cost of moneys.

"Electric Power Companies: How to Pay More for Less," Novick, S., October 1976, 12 pages, DOE Accession Number: 175258, EDB-77:012688. The author states that the problems of the power industry began in the 1960s when electric power rates stopped their 70-year decline and began to rise faster than the prices of other goods and services. The reasons, he notes, for the reverse are not entirely clear. However, they are due in part to the incentives toward waste built into the power industry seem finally to have overcome the increases in efficiency in generating new plants. In addition, for many companies and the industry as a whole, the efficiency of the newest plants measured in terms of "heat rate," began to decline. Interest rates and construction costs, critical cost elements for the power industry, began their slow rise after World War II, and by the 1960s these, too, seemed to be overcoming the efforts of utilities to maintain the growth of their earnings. The author writes that it is not clear how long customers will continue to pay the increased costs of electric power even though conservation is being practiced.

"What Happened to the Nuclear Plant Program in 1975," Olds, F.C., April 1976, 85 pages, DOE Accession Number: 110817, INS-76:013145, EDB-76:047984. The continuation of postponements and cancellations of nuclear power in 1975 indicates the possibility of a shortage of installed capacity of all kinds in the early 1980s. The author notes that the years 1969 and 1970 saw only 21 plants, totaling 21,469 MW, put on order. Five of them were later postponed. He wonders what will be the outcome as a result of slip in scheduling. Costs are likely to continue to grow at a rapid rate as the proposed projects are delayed.

National Technical Information Service (NTIS) Database

The NTIS database consists of government-sponsored research, development, and engineering plus analyses prepared by Federal agencies, their contractors or grantees.

"Study of Factors Inhibiting Effective Use of Domestic Nuclear Power," U.S. Department of Energy, February 21, 1978, 32 pages, NTIS Accession Number: 689375, TID-28694. The sharp reduction in orders for nuclear power plants since the mid-1970s lead to serious concern over the nation's ability to bring new capacity on line five or more years away. Furthermore, nuclear plants under construction and in the licensing process are being delayed and costs are escalating. These trends, if they continue, could have major adverse impact on the costs of electricity. The following growth inhibiting factors are examined: (1) uncertainty over government acceptance of nuclear power, (2) long and uncertain lead times, (3) financing difficulties, (4) fuel cycle problems, (5) fuel supply concerns, (6) lack of standardization, (7) conflicting waste transportation, and (10) lack of government support for resolution of technical problems.

"Nuclear Energy Centers: Economic and Environmental Problems," Dollezhal, N.A. and others, 1977, 16 pages, NTIS Accession Number: 606255, IAEA-CN-36/334. The report deals with qualitative and quantitative analysis of factors and problems which arise in the proposed dispersion of sites of nuclear and fuel cycle plants. These problems include the transportation of radioactive nuclear fuel, need for large land areas, water resources, delays in construction and commercial operation, increased costs of labor and other economic and environmental factors. The report recommends ways of decreasing these difficulties.

"World Energy Supply and Demand and the Future of Nuclear Power," Lantzke, U., May 2, 1977 (available only on microfiche), NTIS Accession Number: 606352, IAEA-CN-36/583. Author writes that demand for non-fossil energy sources will grow substantially in the years to come. The biggest increment to the world's energy supply is expected to come from nuclear power. As a result, the author notes, urgent solutions to problems concerning safety, availability of fuel cycle services, cost escalation and construction delays, etc. will be required.

Electric Power Research Institute Database

The Electric Power Research Institute (EPRI), Palo Alto, California, database covers research on thirteen major categories related to issues in electric power. This database includes research conducted for and by EPRI as well as research conducted by member electric utilities. These reports are available from EPRI's Technical Information Division, telephone number (415) 855-2411.

"Financial Risk from the Operation of a Light Water Reactor Power Plant," General Electric Company, 1983, EPRI Accession Number: 1064600. This project study was a scoping study to develop the technical and financial aspects of risk of investment accompanying the operation of a light water reactor (LWR).

"Evaluation of Costs of Completed Nuclear Units," United Engineers & Constructors, EPRI Accession Number: 209700. The historic costs of nuclear power units show a broad scatter even after correction for size, escalation, and project term. The dominant effect is from initially unplanned accommodation of new regulatory and environmental issues followed by material shortages. The purpose of the study was to identify and understand those variations as they relate to different reactor power units.

"Background Paper on Costs of Nuclear Projects," Lavin Associates, Inc., 1983, EPRI Accession Number: 1142900. The costs of nuclear plants currently under construction are very uncertain say the authors. A portion of the costs are time related and can be estimated from the results of the lead time research. The remaining direct costs are difficult to estimate and projections exhibit wide variation. The purpose of this technical report was to produce a short background paper on the cost outlook for the set of nuclear plants that are currently under construction.

"Evaluation of Nuclear Power Plant Standardization Concepts," Bechtel Group, Inc., 1983, EPRI Accession Number: 1025300. The author(s) used a matrix between stages in the design and construction of the balance of a nuclear power plant and the variables that affect the design and construction. This matrix is intended to identify potential problem areas for standardization of reactor design/plant concepts.

Government Printing Office (GPO) Publications Reference Files

The GPO publications reference files (Washington, D.C.) indexes public documents currently for sale by the Superintendent of Documents, U.S. Government Printing Office, as well as forthcoming and recently out-of-print publications. These publications are produced by the legislative and executive branches of the U.S. Federal Government.

"Nuclear Power Regulation," Nikodem, Zdenek D., DOE, 1980, 230 pages, GPO Number: 8005625, E 3.2:P 75/v.10. This report examines the programs for regulating the safety, design, and operation of domestic nuclear power plants. The first part describes the Federal and state regulatory procedures, and the second part examines the effects of nuclear safety regulations on the planning and construction lead time for nuclear power plants, the cost of nuclear power, and the decision to invest in nuclear power.

Energyline Database

"America's Electric Utilities: Under Siege and in Transition," Fenn, Scott A. and others, Investor Responsibility Research Center Report, 1983, 114 pages, Energyline

Report Number: 142761, 83-024821. The author(s) report that economic and political pressures facing the U.S. electric utility industry are forcing this sector to re-examine its business and operating practices. High inflation and interest rates, increasing construction costs, and lagging technological innovations are some of the economic constraints plaguing the industry and reducing cash flow and revenues. The author(s) say that diminished electric growth and deteriorating financial conditions now characterize the utility industry as a result of these combined factors.

"A Case Not Proven," MacKerron, Gordon, University of Sussex, United Kingdom, January 13, 1983, 76 pages, Energyline Report Number: 141753, 83-023861. The Central Electricity Generating Board believes that the Sizewell nuclear plant at Suffolk, will reduce the costs of electricity to consumers. The author argues that project costs associated with delays and risks of new technology have not been included in the overall benefits cited by the Central Electricity Generating Board.

"Had Construction Not Stopped," Kline, Robert V., Harvard University, Energy Systems and Policy, 1982, 213 pages, Energyline Report Number: 139102, 83-021331. The direct financial consequences of the slowdown of the nuclear power program in New England in the 1970s are reviewed. The author believes that if plans created in the early 1970s had been carried through, electricity costs for 1981–1990 would be cheaper than they are presently and electricity prices would rise less in the coming decade.

"The True Cost of Nuclear Energy: A Special Report," *Ecologist*, 1981, page 253, 40 pages, Energyline Report Number: 135934, 82-023305. The United Kingdom's nuclear power program is considered and its potential assessed. After a review of all considerations it appears that nuclear power costs are rising while the cost of coal remains steady. The report concludes that nuclear power is uneconomical compared to electricity generated by fossil fuels.

"Nuclear Energy in the 80s," Franklin, N.L., Nuclear Power Company, Ltd., United Kingdom, September 1980, 120 pages. Energyline Report Number: 130726, 81-023344. Author writes that world uranium demand in the next 20 years is likely to be substantially less than initially projected due to cancellations and delays in the construction of nuclear power plants. As a result of the lessening demand for uranium, author believes that the price of the fuel is likely to rise due to uncertainties about future markets.

"Extra Cost of Nuclear Power Plants Arising from Conditions Imposed by Authorities," Reiger, A.W., Osterreichische Elektrizitatswirtschafts AG, Austria, December 1980, 616 pages (in German). In this survey report, the author states that nuclear power plants under construction are almost always subjected to additional regulations by public authorities that result in design modifications and increased construction costs.

"The Crossroads of Nuclear Energy," Hogerton, John F., S.M. Stoller Corporation, December 1979, 30 pages, Energyline Report Number: 125028, 80-022932. It has been estimated that installed nuclear capacity in 1990 would be 400,000 MW. However, more recent projections indicate that nuclear capacity will not exceed 140,000 MW. Factors causing the setback include: reduced electricity growth, utility financial constraints, long nuclear lead times, and nuclear power uncertainties.

"Nuclear Lead Times: Are They Too Long," Greenhalgh, Geoffrey, April 1980, 22 pages. Energyline Report Number: 124383, 80-022322. Nuclear power plant orders have fallen from the peak year of 1973 to nearly zero by 1975 and with cancellations even below zero during 1975–1978. Although orders for fossil-fired power plants have also declined dramatically, electricity generation in the U.S. is steadily increasing. The nuclear ordering rate decrease is mirrored in the increasing borrowing rate, delays in construction, increasing final investment costs by as much as 44 percent.

"The First Decade—TVA's First Ten Years of Nuclear Power Plant Design and Construction Experience," Willis, William, TVA Office of Engineering Design and

Construction, April 1978, 101 pages, Energyline Report Number: 121207, 79-024287. Experience gains in TVA's first decade of designing and construction nuclear power plants are reviewed. Since 1966, TVA has completed a three-unit reactor with five more plants in various stages of construction. Assumptions used for estimating capital costs and project schedules for the first TVA units are analyzed. The evolution of the nuclear power situation in the 1970s is discussed. Despite tremendous cost rises, nuclear plants are held to be the most economically attractive option. Regulatory code, equipment market, and cost trends are projected.

"Tennessee Valley Authority Can Improve Estimates and Should Reassess Reserve Requirements for Nuclear Power Plants," GAO Report, March 22, 1979, Energyline Report Number: 119672, 79-022818. TVA's cost estimates for the planned Hartsville, Phipps Bend, and Yellow Creek nuclear power plants in Tennessee and Mississippi are understated by several hundred million dollars each because of specific excluded costs and unrealistic construction schedules.

"Cost Escalation in Nuclear Power," Montgomery, W.D. and Quirk, J.P., California Institute of Technology, January 1978, Energyline Report Number: 119125, 79-022296. A historical review overview of the development of the nuclear power industry and of cost escalation within the industry. The escalation of capital costs of nuclear power plant from 1960–1978 is examined. Included is new data on regulatory delays in the licensing process and the effects of such delays on capital costs of nuclear power plants. It is concluded that nuclear capital costs have escalated more rapidly than either the GNP or the construction price index. Prior to 1970, cost increases were related to bottleneck problems in the nuclear construction and supplying industries, and to the regulatory process. After 1970, generic changes introduced into the licensing process were responsible for most of the cost increases.

"Nuclear Power Plant Lead-Times," Lester, Richard K., MIT, November 1978, Energyline Report Number: 118657, 79-021852. The length and unpredictability of lead times associated with the planning, design, and construction of nuclear power plants are threatening the nuclear power industry.

"Delays in Construction of Nuclear Power Plants," Rad, Parviz F., Clemson University, January 1979, 33 pages, Energyline Report Number: 118074, 79-021293. Interviews were conducted with design and construction personnel of TVA and the Duke Power Company in an effort to identify those areas of nuclear power plant R&D and construction that will facilitate shortening overall construction time. Results indicate that the most important items were overall coordination, equipment availability, design-construct lead time, and handling of design changes.

"Electricity Prices for the Next Decade," Stelzer, Irwin M., National Economic Research Association, January 1975, Energyline Report Number: 101432, 75-005954. Electricity prices are projected to exceed the general level of inflation during the next ten years mainly due to capital costs. Cancellation in construction plans is dramatic: 20.4 percent of megawatts from nuclear plant proposals has been dropped. Cancellations are due to slowdown in the growth rate of electric utilities, rapid escalation in cost of nuclear plant construction, and severe capital shortages.

"Construction Delays Pinpointed by MIT Survey," *Electric Light & Power Magazine*, January 1975, Energyline Report Number: 101284, 75-005306. Average delay in starting fossil-fuel power plant construction is only one month, and the average overrun construction is only 1.7 months. Average duration of construction is 36 months. Most delays of fossil plants are associated with labor problems, especially labor shortages. Average duration of construction for nuclear plants was 60 months, with an average overrun of nearly 16 months. Construction permits, environmental, and labor problems contributed significantly to delays. Cost of nuclear construction delays averaged $148,000/day compared with $70,000/day for fossil plant delays.

Appendix A

"Information Up-Date"
Chemical Bank
55 Water Street
New York, NY 10041
1-800-233-2113
212-820-5007 (inside New York State)
Weekly news and status reports on Chemical Bank's lawsuits against WPPSS and others. Use special toll-free telephone line to get latest information.

Kenneth L. Dowd
Sr. Vice President–Trust Dept.
Chemical Bank
55 Water Street
New York, NY 10041
212-310-6161
Participant in Chemical's closed-circuit TV program with bondholders. Contact him for additional information about Chemical's WNP 4 and 5 settlement activities and plans.

Michael Mines
Attorney–Chemical Bank
900 Fourth Avenue
Seattle, WA 98164
206-292-9988
Attorney for Chemical Bank's lawsuit against WPPSS and based in Seattle. Also participated in Chemical Bank's bondholder closed-circuit TV program.

William H. Berls
Sr. Vice President–Trust Dept.
Chemical Bank
55 Water Street
New York, NY 10041
212-310-6161
Chemical's leading spokesperson to the bondholders and the top person to contact for news and information.

Class Action
Graham & Dunn
1301 Fifth Avenue
Seattle, WA 98101
206-624-8300
Attorney for a bondholders' class action suit against WPPSS. Contact for additional information and progress.

Class Action
J. Porter Kelley
1117 Norton Building
Seattle, WA 98104
206-682-3840
Attorney for another bondholders' class action suit against WPPSS. Contact for additional information and progress.

Cyrus Noe
Bondwatch XIV
Box 9288 Queen Ann Station
Seattle, WA 98109
206-283-4811
Publisher of "Bondwatch XIV," an excellent news report of what is happening on all WPPSS lawsuits. Cost: $120 per year.

William "Bill" Newgent
332 Cameron Road
Port Angeles, WA 98362
206-452-4595
Highly informed, knowledgeable WNP 4 and 5 bondholder who has been instrumental in organizing bondholders political action activities. Although he no longer writes a monthly newsletter on WPPSS settlement and lawsuit activities, he continues to remain active and helpful in efforts to bring about a bondholders settlement.

Securities Exchange Commission
Director
450 5th Street, N.W.
Washington, D.C. 20549
202-272-3100

Contact periodically to find what efforts they are making to see that a fair settlement is made on all WNP 4 and 5 bonds.

Supreme Court of Washington
State of Washington
Temple of Justice
Olympia, WA 98504
206-753-5111

Contact periodically to find what efforts they are making to see that a fair settlement is made on all WNP 4 and 5 bonds.

WPPSS
P.O. Box 968
3000 George Washington Way
Richland, WA 99352
509-372-5000

Should you care to write for whatever reasons.

National WPPSS Bondholders Committee
P.O. Box 4634
Hialeah, Florida 33014

Membership information and news on the activities of Chemical Bank sponsored national bondholders committee.

Index